T0184905

Communications in Computer and Information Science   1069

Commenced Publication in 2007
Founding and Former Series Editors:
Phoebe Chen, Alfredo Cuzzocrea, Xiaoyong Du, Orhun Kara, Ting Liu,
Krishna M. Sivalingam, Dominik Ślęzak, Takashi Washio, and Xiaokang Yang

Editorial Board Members

Simone Diniz Junqueira Barbosa
  Pontifical Catholic University of Rio de Janeiro (PUC-Rio),
  Rio de Janeiro, Brazil

Joaquim Filipe
  Polytechnic Institute of Setúbal, Setúbal, Portugal

Ashish Ghosh
  Indian Statistical Institute, Kolkata, India

Igor Kotenko
  St. Petersburg Institute for Informatics and Automation of the Russian
  Academy of Sciences, St. Petersburg, Russia

Junsong Yuan
  University at Buffalo, The State University of New York, Buffalo, NY, USA

Lizhu Zhou
  Tsinghua University, Beijing, China

More information about this series at http://www.springer.com/series/7899

Xiaoming Sun · Kun He · Xiaoyun Chen (Eds.)

Theoretical Computer Science

37th National Conference, NCTCS 2019
Lanzhou, China, August 2–4, 2019
Revised Selected Papers

 Springer

Editors
Xiaoming Sun
Institute of Computing Technology
Chinese Academy of Sciences
Beijing, China

Kun He
Huazhong University of Science
and Technology
Wuhan, China

Xiaoyun Chen
Lanzhou University
Lanzhou, China

ISSN 1865-0929   ISSN 1865-0937 (electronic)
Communications in Computer and Information Science
ISBN 978-981-15-0104-3   ISBN 978-981-15-0105-0 (eBook)
https://doi.org/10.1007/978-981-15-0105-0

© Springer Nature Singapore Pte Ltd. 2019
This work is subject to copyright. All rights are reserved by the Publisher, whether the whole or part of the material is concerned, specifically the rights of translation, reprinting, reuse of illustrations, recitation, broadcasting, reproduction on microfilms or in any other physical way, and transmission or information storage and retrieval, electronic adaptation, computer software, or by similar or dissimilar methodology now known or hereafter developed.
The use of general descriptive names, registered names, trademarks, service marks, etc. in this publication does not imply, even in the absence of a specific statement, that such names are exempt from the relevant protective laws and regulations and therefore free for general use.
The publisher, the authors and the editors are safe to assume that the advice and information in this book are believed to be true and accurate at the date of publication. Neither the publisher nor the authors or the editors give a warranty, expressed or implied, with respect to the material contained herein or for any errors or omissions that may have been made. The publisher remains neutral with regard to jurisdictional claims in published maps and institutional affiliations.

This Springer imprint is published by the registered company Springer Nature Singapore Pte Ltd.
The registered company address is: 152 Beach Road, #21-01/04 Gateway East, Singapore 189721, Singapore

Preface

The National Conference of Theoretical Computer Science (NCTCS) is the main academic activity in the area of theoretical computer science in China. To date, NCTCS has been successfully held 36 times in over 20 cities. It provides a platform for researchers in theoretical computer science or related areas to exchange ideas and start cooperations.

This volume contains the papers presented at the 37th National Conference of Theoretical Computer Science (NCTCS 2019), held during August 2–4, 2019, in Lanzhou, China. Sponsored by the China Computer Forum (CCF), NCTCS 2019 was hosted by the CCF Theoretical Computer Science Committee and School of Information Science & Engineering at Lanzhou University (LZU).

NCTCS 2019 received 28 English submissions in the area of algorithms and complexity, data science and machine learning theory, parallel and distributed computing, and computational models. Each of the 28 submissions was reviewed by at least three Program Committee members. The committee decided to accept 11 papers that are included in these proceedings published by Springer's *Communications in Computer and Information Science* (CCIS) series.

NCTCS 2019 invited well-reputed researchers in the field of theoretical computer science to give keynote speeches, carry out a wide range of academic activities, and introduce recent advanced research results. We had seven invited plenary speakers at NCTCS 2019: Dingzhu Du (The University of Texas at Dallas, USA), Hai Jin (Huazhong University of Science and Technology, China), Ke Yi (Hong Kong University of Science and Technology, China), Yitong Yin (Nanjing University, China), Pinyan Lu (Shanghai University of Finance and Economics, China), Wenkang Weng (Southern University of Science and Technology, China), and Xinwang Liu (National University of Defense Technology, China). We express our sincere thanks to them for their contributions to the conference and proceedings.

We would like to thank the Program Committee members and external reviewers for their hard work in reviewing and selecting papers. We are also very grateful to all the editors at Springer and the local organization chairs for their hard work in the preparation of the conference.

September 2019

Xiaoming Sun
Kun He
Xiaoyun Chen

Organization

General Chairs

Dingzhu Du   The University of Texas at Dallas, USA
Lian Li   Hefei University of Technology, China

Program Chairs

Xiaoming Sun   Chinese Academy of Sciences, Institute of Computing
     Technology, China
Kun He   Huazhong University of Science and Technology,
     China
Xiaoyun Chen   Lanzhou University, China

Organization Chairs

Xiaoyun Chen   Lanzhou University, China
Zhixin Ma   Lanzhou University, China

PC Members

Xiaojuan Ban   University of Science and Technology Beijing, China
Xiaohui Bei   Nanyang Technological University, Singapore
Kerong Ben   Naval University of Engineering, China
Jin-Yi Cai   University of Wisconsin-Madison, USA
Zhiping Cai   National University of Defense Technology, China
Yixin Cao   The Hong Kong Polytechnic University, SAR China
Yongzhi Cao   Peking University, China
Juan Chen   National University of Defense Technology, China
Xiaoyun Chen   Lanzhou University, China
Heng Guo   The University of Edinburgh, UK
Jin-Kao Hao   University of Angers, France
Kun He   Huazhong University of Science and Technology,
     China
Zhenyan Ji   Beijing Jiaotong University, China
Haitao Jiang   School of Computer Science and Technology,
     Shandong University, China
Chu-Min Li   Université de Picardie Jules Verne, France
Dongkui Li   School of Information Science and Technology, Baotou
     Normal University, China
Jian Li   Tsinghua University, China
Longjie Li   Lanzhou University, China

Lvzhou Li	Zhongshan University, China
Minming Li	City University of Hong Kong, SAR China
Zhanshan Li	Jilin University, China
Peiqiang Liu	School of Computer Science and Technology, Shandong Technology and Business University, China
Qiang Liu	College of Computer, National University of Defense Technology, China
Tian Liu	Peking University, China
Xiaoguang Liu	Nankai University, China
Xinwang Liu	School of Computer, National University of Defense Technology, China
Zhendong Liu	Shandong Jianzhu University, China
Shuai Lu	Jilin University, China
Rui Mao	Shenzhen University, China
Xinjun Mao	National University of Defense Technology, China
Dantong Ouyang	Jilin University, China
Jianmin Pang	Information and Engineering University, China
Zhiyong Peng	State Key Laboratory of Software Engineering, Wuhan University, China
Zhengwei Qi	Shanghai Jiao Tong University, China
Junyan Qian	Guilin University of Electronic Technology, China
Jiaohua Qin	Hunan City University, China
Kaile Su	Griffith University, Australia
Xiaoming Sun	Institute of Computing Technology, Chinese Academy of Sciences, China
Chang Tang	China University of Geosciences, China
Gang Wang	Nankai-Baidu Joint Laboratory, College of Information Technical Science, Nankai University, China
Jianxin Wang	Central South University, China
Liwei Wang	Wuhan University, China
Zihe Wang	Shanghai University of Finance and Economics, China
Jigang Wu	Guangdong University of Technology, China
Zhengjun Xi	Shaanxi Normal University, China
Mingji Xia	Institute of Software, Chinese Academy of Sciences, China
Meihua Xiao	East China Jiaotong University, China
Mingyu Xiao	University of Electronic Science and Technology of China, China
Jinyun Xue	Provincial Key Laboratory of High-Performance Computing, Jiangxi Normal University, China
Yan Yang	School of Information Science and Technology, Southwest Jiaotong University, China
Jianping Yin	National University of Defense Technology, China
Yitong Yin	Nanjing University, China
Mengting Yuan	Wuhan University, China

Defu Zhang	Xiamen University, China
Jialin Zhang	Chinese Academy of Sciences, China
Jianming Zhang	Changsha University of Science and Technology, China
Peng Zhang	School of Computer Science and Technology, Shandong University, China
Yong Zhang	Shenzhen Institutes of Advanced Technology, Chinese Academy of Sciences, China
Zhao Zhang	Zhejiang Normal University, China
Yujun Zheng	Zhejiang University of Technology, China
En Zhu	National University of Defense Technology, China

Abstracts for Invited Talks

Nonsubmodular Optimization and Iterated Sandwich Method

Dingzhu Du

The University of Texas at Dallas, USA
dzdu@utdallas.edu

Abstract. In study of social networks and machine learning, such as viral marking of online game, we may meet nonsubmodular optimization problems. The sandwich method is one of popular approaches to deal with such problems. In this talk, we would like to introduce a new development about this method, the iterated sandwich method, and analysis on the computational complexity and the performance of the iterated sandwich method. This talk is based on a recent research work of research group in the Data Communication and Data Management Lab at University of Texas at Dallas.

Challenges and Practices on Efficient Graph Computing

Hai Jin

Huazhong University of Science and Technology, China
hjin@hust.edu.cn

Abstract. Graph computing is a very challenging task for big data. Apart from a great deal of research relevant to graph computing algorithms, there is a growing attention on investigating efficient graph computing from the aspect of computer architecture. In this talk I would like to investigate the challenges of processing graph computing for the current computer architecture, and discuss the intrinsic properties and computational requirements for graph computing. Then I would like to introduce some possible research directions and approaches to address for efficient graph computing. My team's research progress in efficient graph computing accelerator is introduced in the end.

Join Algorithms in the BSP

Ke Yi

Hong Kong University of Science and Technology, China
yike@cse.ust.hk

Abstract. Many modern distributed data systems adopt a main memory based, shared-nothing, synchronous architecture, exemplified by systems like MapReduce and Spark. These systems can be abstracted by the BSP model, and there has been a strong revived interest in designing BSP algorithms for handling large amounts of data. In this talk, I will present an overview of the recent results on BSP algorithms for multi-way joins (a.k.a., conjunctive queries), a central problem in database theory and systems.

Sampling and Counting for Big Data

Yitong Yin

Nanjing University, China
yinyt@nju.edu.cn

Abstract. Random sampling and approximate counting are classic topics for randomized algorithm. According to the famous Jerrum-Valiant-Vazirani theorem, for all self-reducible problems, random sampling and approximate counting are computationally equivalent for polynomial time probabilistic Turing machine. Random sampling methods like MCMC (Markov chain Monte Carlo) are used so as to implement approximate counting and probabilistic inference, and sampling is import in the area of randomized algorithms and probabilistic graphical model theory. However, these classic algorithms and reduction rely heavily on the sequence and globality of computation as well as the static inputs, and they meet challenges for big data computation. This talk will include the following topics: parallel or distributed sampling algorithm, computational equivalence of random sampling and approximate counting for distributed computing, and sampling algorithm with dynamic inputs.

Approximate Counting via Correlation Decay

Pinyan Lu

School of Information Management and Engineering, Shanghai University
of Finance and Economics, Shanghai, China
Lu.pinyan@mail.shufe.edu.cn

Abstract. In this talk, I will survey some recent development of approximate counting algorithms based on correlation decay technique. Unlike the previous major approximate counting approach based on sampling such as Markov Chain Monte Carlo (MCMC), correlation decay based approach can give deterministic fully polynomial-time approximation scheme (FPTAS) for a number of counting problems. The algorithms have applications in statistical physics, machine learning, stochastic optimization and so on.

Quantum Computing for the Near Future

Wenkang Weng

Southern University of Science and Technology, China
yung@sustech.edu.cn

Abstract. In the near future, it is possible that quantum devices with 50 or more high-quality qubits can be engineered. On one hand, these quantum devices could potentially perform specific computational tasks that cannot be simulated efficiently by classical computers. On the other hand, the number of qubits would not be enough for implementing textbook quantum algorithms. An immediate question is how one might exploit these near-term quantum devices for really useful tasks? In addition, one may also expect that these powerful quantum devices are accessible only through cloud services over the internet, which imposes the question of how might one verify the server, behind the internet, does own a quantum computer instead of a classical simulator? In this talk, I will share my thoughts over these questions based on my recent works.

Incomplete Multi-view Clustering Algorithms

Xinwang Liu

National University of Defense Technology, China
1022xinwang.liu@gmail.com

Abstract. We focus on multi-view clustering in this talk. We propose a matrix norm regularization based multi-modal clustering algorithm so as to reduce the redundancy and enhance the diversity. Second, we propose an incomplete multi-modal classification and clustering algorithm for learning problems with incomplete multi-modal. Third, we propose a noisy multi-modal classification and clustering algorithm for learning problems with noisy modals.

Contents

Algorithms and Complexity

Cooperation on the Monte Carlo Rule: Prisoner's Dilemma Game
on the Grid 3
   Jiadong Wu and Chengye Zhao

Special Frequency Quadrilaterals and an Application 16
   Yong Wang

Data Science and Machine Learning Theory

Sampling to Maintain Approximate Probability Distribution
Under Chi-Square Test 29
   Jiaoyun Yang, Junda Wang, Wenjuan Cheng, and Lian Li

Adjusting the Inheritance of Topic for Dynamic Document Clustering 46
   Ruizhang Huang, Yingxue Zhu, Yanping Chen, Yue Yang, Weijia Xu,
   Jian Yang, and Yaru Meng

An Improved Proof of the Closure Under Homomorphic Inverse
of FCFL Valued in Lattice-Ordered Monoids 64
   Haihui Wang, Luyao Zhao, and Ping Li

Socially-Attentive Representation Learning for Cold-Start Fraud
Review Detection 76
   Qiaobo Da, Jieren Cheng, Qian Li, and Wentao Zhao

Computational Model

Semi-online Machine Covering on Two Hierarchical Machines
with Known Total Size of Low-Hierarchy Jobs 95
   Man Xiao, Gangxiong Wu, and Weidong Li

A Combined Weighted Concept Lattice for Multi-source Semantic
Interoperability of ECG-Ontologies Based on Inclusion Degree
and Information Entropy 109
   Kai Wang

Minimizing the Spread of Rumor Within Budget Constraint
in Online Network 131
   Songsong Mo, Shan Tian, Liwei Wang, and Zhiyong Peng

Quantum Reversible Fuzzy Grammars 150
    Jianhua Jin and Chunquan Li

A Signcryption Scheme Based Learning with Errors over Rings
Without Trapdoor 168
    Zhen Liu, Yi-Liang Han, and Xiao-Yuan Yang

Author Index 181

Algorithms and Complexity

Cooperation on the Monte Carlo Rule: Prisoner's Dilemma Game on the Grid

Jiadong Wu[ID] and Chengye Zhao[⊠][ID]

China Jiliang University, Hangzhou 310018, Zhejiang, China
cyzhao@cjlu.edu.cn

Abstract. In this paper, we investigate the prisoner's dilemma game with monte carlo rule in the view of the idea of the classic monte carlo method on the grid. The monte carlo rule is an organic combination of the current dynamic rules of individual policy adjustment, which not only makes full use of information but also reflects individual's bounded rational behavior and the ambivalence between the pursuit of high returns and high risks. In addition, it also reflects individual's behavioral execution preferences. The implementation of monte carlo rule brings an extremely good result, higher cooperation level and stronger robustness are both achieved by comparing with unconditional imitation rule, replicator dynamics rule and fermi rule. When analyse the equilibrium density of cooperators as a function of the temptation to defect, it appears a smooth transition between the mixed state of coexistence of cooperators and defectors and the pure state of defectors when enhancing the temptation, which can be perfectly characterized by the trigonometric behavior instead of the power-law behavior discovered in the pioneer's work. When discuss the relationship between the temptation to defect and the average returns of cooperators and defectors, it is found that cooperators' average returns is almost a constant throughout the whole temptation parameter ranges while defectors' decreases as the growth of temptation. Additionally, the insensitivity of cooperation level to the initial density of cooperators and the sensitivity to the social population have been both demonstrated.

Keywords: Cooperation · Prisoner's dilemma · Monte carlo rule · Robustness

1 Introduction

The cooperation among selfish individuals [1] such as Kin cooperation [2], Mutually cooperation and Reputation-seeking cooperation [3], which are contrary to natural selection, are widely discovered in human society [4–6], having already become one of the challenges of evolutionary game theory [7–9].

This phenomenon can be characterized by prisoner's dilemma game [10], in which individuals adopt one of two strategies: C (cooperation) or D (defection). A selfish individual would select D as his strategy for the sake of higher returns.

© Springer Nature Singapore Pte Ltd. 2019
X. Sun et al. (Eds.): NCTCS 2019, CCIS 1069, pp. 3–15, 2019.
https://doi.org/10.1007/978-981-15-0105-0_1

Nevertheless, if both sides choose D simultaneously, each of them would get less returns than those acquired for mutual cooperation.

Over the years, scholars often focus exclusively on the promotion of cooperation on different spatial structures [11–13]. However, there are not many researches on the dynamic rules of game individual policy adjustment. Different update rules often lead to different results. For example, Sysi-Aho et al. [14] have modified a more rational update rule on the basis of the research of Harut and Doebeli [15], believing that the individuals have original intelligence in the form of local decision-making rule deciding their strategies. In this rule, individuals suppose their neighbors' strategies retain unchanged and aim at choosing a strategy to maximize their instant returns. This rule results in the density of cooperator at equilibrium which differ tremendously from those resulting from the replicator dynamics rule in the literature [15], and the cooperation can persist throughout the whole temptation parameter ranges. Li et al. [16] adopt the unconditional imitation rule used in the Nowak and May's work [17] to revise the regulation of replication dynamics owing to its briefness and the ability of according with the psychology of most individuals. It is discovered that in some parameter ranges, the performance of inhibiting cooperative behaviors originally turns to promoting. Xia et al. [18] compare the effect on cooperation under unconditional imitation rule, replicator dynamics rule and moran process [19] respectively, discovering that moran process promotes cooperation much more than the others. Those facts affirm the status of update rules in the evolution of cooperative behavior.

In the existing research, the dynamic rules of game individual strategy adjustment are mainly unconditional imitation rule, replicator dynamics rule and fermi rule [20]. However, replicator dynamics and fermi rule can not make full use of the game information while unconditional imitation rule can not tolerate individuals' irrational behavior. An excellent update rule that simulates individual policy adjustment more perfectly needs to be discovered.

In this paper we propose monte carlo rule, which is an organic combination of the current dynamic rules of individual policy adjustment, not only making full use of information but also reflecting individual's bounded rational behavior and ambivalence between the pursuit of high interest and high risk [21,22]. In addition, it also reflects individual's behavioral execution preferences [23,24]. We analyse the effect on cooperative level under monte carlo rule and verify its robustness. Further, we analyse the equilibrium density of cooperator as a function of the temptation to defect and use trigonometric curves to characterize it. It is confirmed that the trigonometric fitting effect is better than the power-law fitting in the pioneer's work [20]. We also investigate the relationship between the temptation to defect and the average returns of cooperators and defectors. Additionally, the insensitivity of cooperation level to the initial density of cooperators and the sensitivity to the social population have been both demonstrated by numerical simulation.

2   Prisoner's Dilemma Game Model with Monte Carlo Rule and Other Three Classic Update Rules

Prisoner's dilemma represents a class of game models that its Nash equilibrium only falls on the non-cooperation. In this game model, each individual can adopt one of two strategies C (cooperation) and D (defection). The returns depend on the strategies of both sides. Enormous temptation forces rational individuals to defect. However, if both sides choose the strategy of D simultaneously, each of them will get less returns than those acquired for mutual cooperation.

Game individuals are located on the nodes of the grid, and the edges indicate the connections between one individual and another. Along the footsteps of Nowak et al. [25], the game returns can be simplified as the matrix below:

	Cooperate	Defect
Cooperate	1	0
Defect	b	0

Where parameter b characterizes the interests of the temptation to defect ($1 < b \leq 2$). The larger the value of b, the greater the temptation of defection to the individuals. For example, if individual i chooses D while his game opponent chooses C, then individual i will get returns $\pi(D, C) = b$ in this game.

Individuals gain returns by playing prisoner's dilemma game with their nearest neighbors. So the returns of individual i in a game round can be expressed as

$$U_i = \sum_{j \in \Omega_i} \pi(s_i, s_j) \tag{1}$$

where Ω_i is the set of neighbors of individual i and s_i, s_j represent the strategies of individual i, j respectively.

During the evolutionary process, each individual adopts one of neighbors' strategies whose returns is more than or equal to himself's in the way of roulette, or just insists the original strategy:

$$P(s_i \leftarrow s_j) = \frac{\omega(i,j)U_j}{\sum_{j \in (i \cup \Omega_i)} \omega(i,j)U_j} \quad, \quad j \in (i \cup \Omega_i) \tag{2}$$

where $P(s_i \leftarrow s_j)$ shows the probability that individual i would imitate j and Ω_i is the set of neighbors of individual i. $\omega(i,j)$ is a characteristic function:

$$\omega(i,j) = \begin{cases} 0 & U_i > U_j, \\ 1 & U_i \leq U_j. \end{cases} \tag{3}$$

In this paper, we also compare monte carlo rule with three classic update rules: unconditional imitation rule, replicator dynamics rule and fermi rule. Here, we briefly outline them:

1. Unconditional imitation rule: during the evolutionary process, each individual would compare his own returns with those of all the neighbors and choose the game strategy with the highest returns as his strategy in next round of game.

2. Replicator dynamics rule: during the evolutionary process, each individual (may be i) randomly choose a neighbor (may be j) for returns comparison. If j's game returns (U_j) is greater than i's (U_i), then individual i would imitate j's strategy with probability p in the next game round:

$$p = \frac{U_j - U_i}{D \cdot \max(k_i, k_j)} \tag{4}$$

where U_i, U_j represent the returns of individual i, j in the previous game round respectively, and k_i, k_j are the number of neighbors they have. Parameter D is the difference between the largest and the smallest parameters in the game matrix (i.e., b).

3. Fermi rule: during the evolutionary process, each individual (may be i) randomly choose a neighbor (may be j) for returns comparison, imitating j's strategy with a certain probability which depends on the difference between the two:

$$p = \frac{1}{1 + e^{(U_i - U_j)/\lambda}} \tag{5}$$

where λ represents the noise effect according to the rationality of individuals.

3   Results and Discussion on Simulation Experiment

3.1   Spatial Structure Plays a Positive Role in Cooperative Behavior in Prisoner's Dilemma Game with Monte carlo Rule on the Grid

For the sake of investigating the influence of spatial structure to cooperative behavior, we apply classical Mean Field Theory [26] which is insensitive to the topology, to preliminary predict the density of cooperators characterized by ρ. Under this circumstance, the average returns of cooperators can be expressed as:

$$U_C = d \cdot \rho \cdot \pi(C, C) + d \cdot (1 - \rho) \cdot \pi(C, D) = d\rho \tag{6}$$

where d characterizes the average number of neighbors of an individual. In the same way, the average returns of defectors can be expressed as:

$$U_D = d \cdot \rho \cdot \pi(D, C) + d \cdot (1 - \rho) \cdot \pi(D, D) = bd\rho \tag{7}$$

where the b characterizes the temptation to defect. Following the monte carlo rule, we have obtained the differential equation of ρ:

$$\frac{\partial \rho}{\partial t} = (1 - \rho)W(D \leftarrow C) - \rho W(C \leftarrow D)$$

$$= -\rho(1 - \rho)\frac{d \cdot U_D}{d(1 - \rho)U_D + (1 + d\rho)U_C} \tag{8}$$

$$= -\rho(1 - \rho)\frac{1}{1 - \rho + \dfrac{\rho}{b} + \dfrac{1}{bd}}$$

where $W(D \leftarrow C)$ indicates the probability that a defector transforms into a cooperator and $W(C \leftarrow D)$ indicates the probability that a cooperator transforms into a defector. t is the game round.

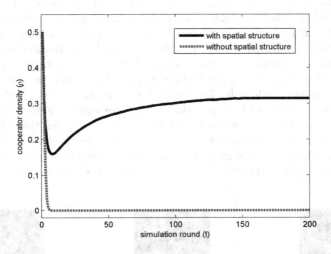

Fig. 1. Numerical simulation of cooperator density under monte carlo rule (solid line) at b = 1.10. The dotted line represents the well-mixed case obtained by Mean Field Theory in the same parameter environment. Those data are simulated on grid of 100×100 where 50% cooperators as well as 50% defectors are randomly distributed at the beginning, and 1000 simulations are averaged in each case.

On the basis of the differential equation, it turns out that as the game progresses, the density of the cooperator (ρ) decreases monotonically until it approaches 0, that is, all the cooperators would go extinct (dotted line in Fig. 1). However, with the spatial structure (e.g., individuals play game on the grid in this paper), cooperators can survive in the form of clusters where they can get support from peers (solid line in Fig. 1), reaffirming the positive role of spatial structure in cooperative behavior [16].

Table 1. A brief comparison of four update rules in this paper.

Update rule	Information utilization	Irrationality
Unconditional imitation	Full	Not exists
Replicator dynamics	Not full	Exists
Fermi	Not full	Exists
Monte carlo	Full	Exists

3.2   The Monte Carlo Rule Can Induce a Highly Cooperative Society and Has Good Robustness

Similar to the unconditional imitation rule, monte carlo rule makes full use of the game information when compared with fermi rule and replicator dynamics rule (see Table 1). That is, during the policy update phase, individuals would collect game information from all the neighbors to determine the most satisfactory strategy in the next game round, instead of just randomly selecting a neighbor to decide whether to imitate or not. This behavior reflects the rigor of individuals. Also, individual's psychology of pursuing return growth is reflected vividly in the monte carlo rule. Therefore, the game system under this rule can support the germination of cooperation to a large extent. Figure 2 shows this fact: again the social population $N = 10000$ and $b = 1.10$, yet there is only one very small cooperative group in the middle of the network at $t = 0$, cooperative behavior can still spread promptly. The middlemost subgraph in Fig. 2 shows the track of function $\rho(t)$, and other subgraphs display the distribution of coopera-

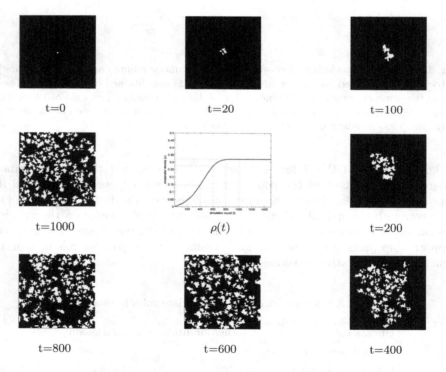

Fig. 2. Snapshots of spatial distribution of cooperators (white boxes) and defectors (black boxes) at $t = 0$, 20, 100, 200, 400, 600, 800 and 1000 when there is only one very small cooperative group in the center of the grid at the initial game moment. The middlemost subgraph shows the track of function $\rho(t)$, and the value of b is still set to be 1.10.

tors and defectors at several important simulate moments. At the beginning, the small cooperative group spreads cooperative behavior in the form of clusters with irregular shapes. At the edge of the clusters, the cooperators resist the temptation of the outside world through the support of peers in the clusters. Owing to the protection of marginal cooperators, the individuals within the clusters are undoubtedly the loyal defenders of cooperation. This model of mutual support enables cooperators to survive in the society. As the game progresses, the rate of propagation continues to increase, reaching the peak at around $t = 500$. Then the rate is slowly reduced to 0, achieving dynamic balance at around $t = 800$. It is clearly that driven by the idea of pursuing progress and full game information, the cooperative behavior quickly spreads throughout the whole network.

Meanwhile, monte carlo rule continues the inclusiveness of fermi rule and moran process for irrational behavior [27]. It considers individual's game returns as his fitness for the society, and the higher the game returns, the more the one can adapt to the society. In general, monte carlo rule is an organic combination of the current dynamic rules of individual policy adjustment, so that it not only makes full use of information, but also reflects individual's bounded rational behavior and ambivalence between the pursuit of high interest and high risk.

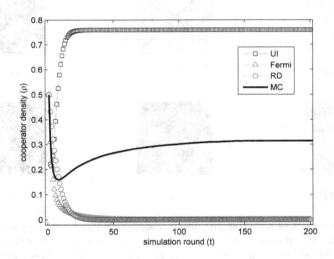

Fig. 3. Numerical simulation of cooperator density under monte carlo rule (solid line), replicator dynamics rule (circles), fermi rule (triangles) and unconditional imitation rule (squares) on the grid at $b = 1.10$. 1000 simulations are averaged in each case.

Further, we investigate the role of update rules in the prisoner's dilemma game on the grid. We takes $\lambda = 0.0625$ in fermi rule for the reason that, in this case the track of $\rho(t)$ is closest to those under monte carlo rule with the guidance of Mean Field Theory. This setting makes it fairer to compare the effects on cooperation level. Since individuals cannot judge the priority of the two strategies at the beginning, their initial strategies are all based on coin toss. That is, cooperators and defectors occupy 50% of the grid, respectively. The results are showed in Fig. 3. During the PD game, the ρ tends to be stable promptly. We observe that the aggregate cooperation level between individuals is largely elevated under unconditional imitation rule or monte carlo rule, when compared to fermi rule and replicator dynamics. It is clearly that update rules play an important role in the evolutionary theory [19].

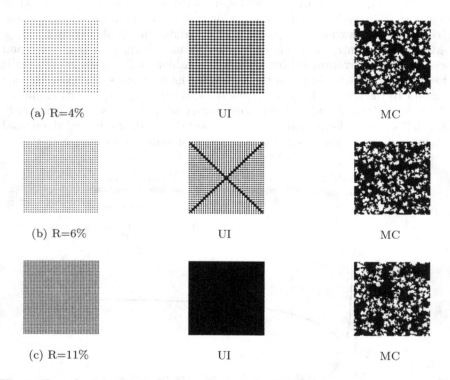

Fig. 4. The influence of defective invasions on society. White boxes represent coopera-tors and black boxes represent defectors. Each row corresponds to the different number of defectors (R), which are around 4%, 6%, 11% of the society, respectively. The first column shows the strategic distribution when invasion occurs, the second column shows the game equilibrium results under the unconditional imitation rule, and the third col-umn shows the game equilibrium results under the monte carlo rule. N is set to be 10000 and b is set to be 1.10.

Although cooperation level under unconditional imitation rule is higher than that of monte carlo rule, its mechanism of just imitating the best without any hesitation makes individuals no longer consider any risks behind the high-yields, so there is a reason to believe that it is not of robustness. To confirm this fact, we consider an extreme situation. All the individuals on the grid are cooperators, and at some point, a group deliberately change its game strategy (defect). After that, we simulate enough time steps for accommodation of the defective invasion. The results are shown in Fig. 4. Each row corresponds to the different number of defectors (R), which are around 4%, 6%, 11% of the society, respectively. The first column shows the strategic distribution when the invasion occurs, the second column shows the game equilibrium results (we say the game reaches equilibrium when the density of the cooperator almost no longer changes with the game round) under the unconditional imitation rule, and the third column shows the game equilibrium results under the monte carlo rule. As the invasion of the defection, defectors' neighbors did not hesitate to imitate the defective strategy for higher returns under unconditional imitation, formed the defective core, which could not update its own strategy, resulting in a significant reduction in cooperation level. However, under monte carlo rule, individuals not only seek for high returns, but also concern the high risks behind the high returns and would make a balance between the two. So no matter how serious the invasion, the cooperation will be balanced over time ($\rho = 0.32$), not extinct. It is overwhelmingly clear that the robustness of monte carlo rule is significantly better than unconditional imitation rule.

3.3 The Cooperator Density Data Refer to a Trigonometric Behavior Under the Monte Carlo Rule

Next, we investigate the fraction of cooperators (ρ) as a function of the temptation parameter (b) under monte carlo rule on the grid since $\rho(b)$ is a pretty meaningful quantity in evolutionary games. The results are showed in Fig. 5(a). The monte carlo rule always performs a cooperative society when temptation is low. As b increases, the balanced pattern has been broke. On the border, the internal support can not conquer the loss caused by defectors, and the cluster begins to collapse layer by layer until a new balance appears. The cooperators go extinct ultimately at $b = 1.31$, for the reason that the huge temptation prevents any form of clusters from overcoming the loss on the border.

According to Szabo's outstanding work, the cooperator density data refer to a power-law behavior in prisoner's dilemma game under the fermi rule on the grid, that is $\rho \propto (b_{cr} - b)^{\beta}$ [20], in which b_{cr} is the threshold of the disappearance of cooperation, once $b > b_{cr}$, then all the cooperators would go extinct. With the guidance of the goodness of fit (R) for nonlinear regression and Root Mean Squared Error ($RMSE$) displayed below:

$$R = 1 - (\sum (y - \hat{y})^2 / \sum y^2)^{1/2} \tag{9}$$

$$RMSE = ((\sum (y - \hat{y})^2)/n)^{1/2} \tag{10}$$

where y represents the original data, $\hat{y}$ is on behalf of the fitting data, and n is the length of y. We find that power-law fitting originally performing very well

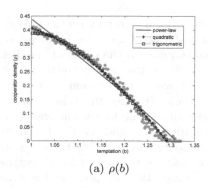
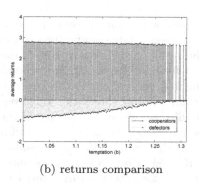

(a) $\rho(b)$ (b) returns comparison

Fig. 5. (a) The effect of power law fitting (solid line), quadratic fitting (pluses) and trigonometric fitting (squares) on the curve $\rho(b)$. For easy comparison, the original data expressed by gray dots are also given in the figure. (b) The relationship between average returns of individuals and the temptation parameter (b). 1000 simulations are averaged in each case.

under the fermi rule is unsatisfactory under the monte carlo rule, which can be seen in Fig. 5(a). The solid line represents the best result of the power-law fitting, where $b_{cr} = 1.31$ and $\beta = 0.923$. After further observing the original data, we choose quadratic and trigonometric curves to fit them. Their expressions can be respectively indicated as:

$$y = ax^2 + bx + c \text{ and } y = A\sin(\omega x + \varphi) + d \qquad (11)$$

The fitting effects are also showed in Fig. 5(a), and the specific fitting value can be seen in Table 2. The pluses represent the results of quadratic fitting where the most optimal parameter are $a = -2.7363$, $b = 4.8644$ and $c = -1.7205$. The squares represent trigonometric's fitting effect where the best parameters are $A = -0.2568$, $w = 7.2462$, $\varphi = -2.5603$ and $d = 0.1314$. The fitting effects of the two are better than that of power-law fitting based on the smaller $RMSE$ and the better R. Furthermore, trigonometric fitting has better results than quadratic fitting, so we say that the cooperator density data refer to a trigonometric behavior under the monte carlo rule on the grid, that is $\rho \propto A\sin(\omega x + \varphi) + d$.

Table 2. The optimal solution of thr three fitting methods.

Fitting method	Root Mean Squared Error	Goodness of fit
Power-law	0.2608	0.9052
Quadratic	0.0168	0.9356
Trigonometric	0.0149	0.9428

3.4 Cooperation Makes Individuals Live Better in Society

We have also observed the relationship between average returns of individuals and the temptation parameter b. The results are showed in Fig. 5(b). Macroscopically, the strategy of cooperation is obviously better than defection due to the better returns. It is worth noting that the average returns of cooperators is insensitive to the temptation parameter b. Furthermore, it is surprised to find that the returns of the defectors decreases as the growth of b, for the reason that excessive defectors gatherings have greatly reduced the success of defection. It is clear that cooperation makes individuals live better in society.

3.5 A Highly Cooperative Society Does Not Depend the Initial Fraction of Cooperators but a Sufficient Social Population

In the investigation above, we choose 50% as the initial density of cooperators ($\rho_0 = 0.5$). In this section we would illustrate the effect of different values of ρ_0 on the cooperators density at equilibrium (ρ). Figure 6(a) shows the real-time fraction of cooperators ($\rho(t)$) for different initial cooperator densities (ρ_0). Those data obtained by taking the average value after 3000 times through the same experiment are simulated on the grid of 100×100 at $b = 1.1$ and $N = 10000$.

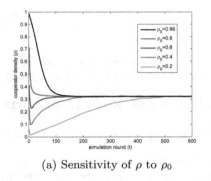

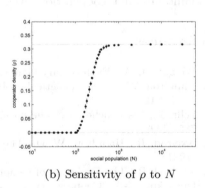

(a) Sensitivity of ρ to ρ_0 (b) Sensitivity of ρ to N

Fig. 6. (a) The numerical simulation results for the density of cooperator (ρ) as the game progresses for different initial density of cooperator (ρ_0) on the grid at $b = 1.10$. From bottom to top, the ρ_0 is set to be 0.2, 0.4, 0.6, 0.8 and 0.99, respectively. (b) The numerical simulation results for the density of cooperator (ρ) at equilibrium for different social population (N) at $b = 1.10$. 3000 simulations are averaged in each case.

As we see, different value of ρ_0 can only appreciably affect the time it takes for the game to reach equilibrium, but do not change the fraction of cooperators at equilibrium. Thus we can say that the fraction of cooperators at equilibrium is insensitive to the initial fraction of cooperators.

Finally, we investigate the effect of the social population (N) on cooperation level. We fix $b = 1.10$ and $\rho_0 = 0.5$, then change social population to obtain corresponding fraction of cooperators at equilibrium. The results are showed in Fig. 6(b). For $N \geq 800$, ρ does not depend on the social population. However, for $N \leq 800$, ρ decreases with smaller N. It is clear that a highly cooperative society depends on a sufficient social population.

4   Conclusions

We introduce a new dynamic rule of game individual strategy adjustment, that is, monte carlo rule and investigate the prisoner's dilemma game under it on the grid. The monte carlo rule not only promotes cooperative behavior, but also has higher robustness when compared with unconditional imitation rule, replicator dynamics rule and fermi rule. Under this rule, spatial structure plays a positive role in cooperative behavior, and the equilibrium density of cooperator as a function of the temptation to defect can be perfectly characterized by a trigonometric behavior instead of the power-law behavior discovered in the pioneer's work under the fermi rule. The society obviously welcomes the cooperation: cooperators can obtain higher and stabler returns than defectors throughout the whole temptation parameter ranges. In addition, the cooperation level is insensitive to the initial density of cooperators but enough social population is needed to maintain a high cooperation level.

References

1. Weibull, J.-W.: Evolutionary Game Theory. MIT Press, Cambridge (1995)
2. Hamilton, W.-D.: The genetical evolution of social behaviour II. J. Theor. Biol. **7**, 17 (1964)
3. Fehr, E., Fischbacher, U.: The nature of human altruism. Nature (London) **425**, 785 (2003)
4. Axelrod, R., Hamilton, W.-D.: The evolution of cooperation. Science **211**, 1390 (1981)
5. Axelrod, R.: The Evolution of Cooperation. Basic Book, New York (1984)
6. Dugatkin, L.-A.: Cooperation Among Animals. Oxford University Press, Oxford (1997)
7. Neumann, J.-V., Morgenstern, O.: Theory of Games and Economic Behaviour. Princeton University Press, Princeton (1944)
8. Smith, J.-M., Price, G.: The logic of animal conflict. Nature **246**, 15 (1973)
9. Fudenberg, D., Levine, D.-K.: The Theory of Learning in Games. The MIT Press, Cambridge (1998)
10. Rapoport, A., Chammah, A.-M.: Prisoner's Dilemma. University of Michigan Press, Ann Arbor (1970)
11. Vukov, J., Szabo, G., Szolnoki, A.: Cooperation in the noisy case: prisoner's dilemma game on two types of regular random graphs. Phy. Rev. E **73**, 067103 (2006)
12. Hauert, C., Szabo, G.: Game theory and physics. Am. J. Phys. **73**, 405 (2005)
13. Tomassini, M., Pestelacci, E., Luthi, L.: Social dilemmas and cooperation in complex networks. Int. J. Modern Phys. C **18**, 1173–1185 (2007)

14. Sysi-Aho, M., Saramaki, J., Kertesz, J., Kaski, K.: Spatial snowdrift game with myopic agents. Eur. Phys. J. B **44**, 129–135 (2005)
15. Hauert, C., Doebeli, M.: Spatial structure often inhibits the evolution of cooperation in the snowdrift game. Nature **428**, 643–646 (1973)
16. Li, P.-P., Ke, J.-H., Lin, Z.-Q.: Cooperative behavior in evolutionary snowdrift games with the unconditional imitation rule on regular lattices. Phys. Rev. E **85**, 021111 (2012)
17. Nowak, M.-A., May, R.-M.: Evolutionary games and spatial chaos. Nature **359**, 826–829 (1992)
18. Xia, C.-Y., Wang, J., Wang, L., Sun, S.-W., Sun, J.-Q., Wang, J.-S.: Role of update dynamics in the collective cooperation on the spatial snowdrift games: beyond unconditional imitation and replicator dynamics. Chaos Solitons Fractals **45**, 1239–1245 (2012)
19. Pap, M.: The Statistical Process of Evolutionary Theory. Clarendon Press, Oxford (1962)
20. Szabo, G., Toke, C.: Evolutionary prisoner's dilemma game on a square lattice. Phys. Rev. E **58**, 69–73 (1998)
21. Huang, D.-X.: Game analysis of demand side evolution of green building based on revenue-risk. J. Civil Eng. (Chin. Ed.) **50**(2), 110–118 (2017)
22. He, L.-J., Yang, L.-P.: Travel agency income nash equilibrium model influenced by risk preference. Guangdong Univ. Technol. (Chin. Ed.) **36**(3), 56–67 (2019)
23. Coultas, J.-C.: When in Rome: an evolutionary perspective on conformity. Group Process. Intergroup Relat. **7**(4), 317–331 (2004)
24. Efferson, C., Lalive, R., Richerson, P.-J., et al.: Conformists and mavericks: the empirics of frequency-dependent cultural transmission. Evol. Hum. Behav. **29**(1), 56–64 (2008)
25. Nowak, M.-A., May, R.-M.: The spatial dilemmas of evolution. Int. J. Bifurcation Chaos Appl. Sci. Eng. **3**, 35 (1993)
26. Wu, Z.-X.: Complex Network and Evolutionary Game Research on It (Chinese edition). Lanzhou University, Lanzhou (2007)
27. Lieberman, E., Hauert, C., Nowak, M.-A.: Evolutionary dynamics on graphs. Nature **433**(7023), 312–316 (2005)

Special Frequency Quadrilaterals
and an Application

Yong Wang$^{(\boxtimes)}$ (iD)

North China Electric Power University, Beijing 102206, China
yongwang@ncepu.edu.cn

Abstract. Given a quadrilateral $ABCD$ in K_n and the distances of edges, the special frequency quadrilaterals are derived as two of the three sum distances $d(A,B) + d(C,D)$, $d(A,C) + d(B,D)$, and $d(A,D) + d(B,C)$ are equal. A probability model formulated based on the special frequency quadrilaterals implies the edges in the optimal Hamiltonian cycle are different from the other edges in K_n. Christofides proposed a $\frac{3}{2}$-approximation algorithm for metric traveling salesman problem (TSP) that runs in $O(n^3)$ time. Cornuejols and Nemhauser constructed a family of graphs where the performance ratio of Christofides algorithm is exactly $\frac{3}{2}$ in the worst case. We apply the special frequency quadrilaterals to the family of metric TSP instances for cutting the useless edges. In the end, the complex graph is reduced to a simple graph where the optimal Hamiltonian cycle can be detected in $O(n)$ time, where $n \geq 4$ is the number of vertices in the graph.

Keywords: Traveling salesman problem ·
Special frequency quadrilateral · Heuristic algorithm

1 Introduction

Traveling salesman problem (TSP) is one of well-known NP-hard problems in combinatorial optimization [1]. The aim is to find an optimal Hamiltonian cycle (OHC) in a given complete graph K_n. Since many complex industrial problems have close relationships with TSP, the algorithms for resolving TSP or reducing its complexity will be widely accepted in engineering applications. Karp [2] has shown that TSP is NP-complete. It implies that there are no polynomial-time algorithms for TSP unless $P = NP$. In the worst case, the computation time of exact algorithms is $O(a^n)$ for some $a > 1$ for TSP. For example, Held and Karp [3], and independently Bellman [4] designed a dynamic programming approach that consumed $O(n^2 2^n)$ time. Integer programming techniques such as either branch and bound [5,6] or cutting-plane [7,8] have been able to solve TSP examples on thousand points. For large TSP, the computation times of these exact algorithms is hard to reduce.

The authors acknowledge the funds supported by the Fundamental Research Funds for the Central Universities (No. 2018MS039 and No. 2018ZD09).

© Springer Nature Singapore Pte Ltd. 2019
X. Sun et al. (Eds.): NCTCS 2019, CCIS 1069, pp. 16–26, 2019.
https://doi.org/10.1007/978-981-15-0105-0_2

On the other hand, the computation times of approximation algorithms and heuristics have been significantly decreased. For example, the MST-based algorithm [9] and Christofides' algorithm [10] are able to find the 2-approximation and $\frac{3}{2}$-approximation in time $O(n^2)$ and $O(n^3)$, respectively, for metric TSP. Moreover, Mömke and Svensson [11] gave a 1.461-approximation algorithm for metric graphs with respect to the Held-Karp lower bound. In most cases, the Lin-Kernighan heuristics (LKH) can compute the "high quality" solutions within 2% of the optimum in nearly $O(n^{2.2})$ time [12].

In recent years, researchers have designed polynomial-time algorithms to resolve the TSP on sparse graphs. In sparse graphs, the number of Hamiltonian cycles (HC) is greatly reduced. For example, Sharir and Welzl [13] proved that in a sparse graph of average degree d, the number of HCs is less than $e^*(\frac{d}{2})^n$, where e^* is the base of natural logarithm. Heidi [14] gave a lower bound $(\frac{d}{2})^n$ for a graph of average degree d. In addition, Björklund [15] proved that a TSP on bounded degree graphs can be solved in time $O(2 - \epsilon)^n$, where ϵ depends on the maximum degree of a vertex. Given a TSP on cubic graphs, Eppstein [16] introduced an exact algorithm with running time $O(1.260)^n$. This run time was improved by Liśewicz and Schuster [17] to $O(1.253)^n$. Aggarwal, Garg and Gupta [18] and independently Boyd, Sitters, Van der Ster and Stougie [19] gave two $\frac{4}{3}$-approximation algorithms to solve the TSP on metric cubic graphs. For TSP on cubic connected graphs, Correa, Larré and Soto [20] proved that the approximation threshold is strictly below $\frac{4}{3}$. For the general bounded-genus graphs, Borradaile, Demaine and Tazari [21] gave a polynomial-time approximation scheme for TSP. In the case of asymmetric TSP, Gharan and Saberi [22] designed the constant factor approximation algorithms for TSP. For planar graphs with bounded genus, the constant factor is $22.51(1 + \frac{1}{n})$. Thus, whether one is trying to find exact solutions or approximate solutions to the TSP, one can has a variety of more efficient algorithms available if one can reduce a given TSP to finding an OHC in a sparse graph.

In the edges elimination research, Jonker and Volgenant [23] found many useless edges out of OHC based on $2 - opt$ move. After these edges were cut, the computation time of branch-and-bound for certain TSP instances was reduced to half. Hougardy and Schroeder [24] cut the useless edges based on $3 - opt$ move. Their combinatorial algorithm cuts more useless edges for TSP, and the computation time of Concorde was decreased by more than 11 times for certain TSP instances in $TSPLIB$. Different from the above research, the authors eliminate the edges out of OHC for TSP according to frequencies of edges computed with frequency quadrilaterals [25, 26] and optimal four-vertex paths [27]. Whether the frequency of an edge is computed with frequency quadrilaterals or it is computed with optimal 4-vertex paths, the frequency of an OHC edge is generally much bigger than that of a common edge. When the minimum frequency of the OHC edges is taken as a frequency threshold to trim the other edges with small frequencies, the experiments demonstrated that a sparse graph with $O(nlog_2(n))$ edges are obtained for most TSP instances.

In this paper, the special frequency quadrilaterals are computed for a quadrilateral $ABCD$ in K_n where the two of the three distances $d(A,B) + d(C,D)$, $d(A,C) + d(B,D)$, and $d(A,D) + d(B,C)$ are equal. Based on the special frequency quadrilaterals, a probability model is built for the OHC edges. The probability model shows the difference between the OHC edges and the general edges in K_n. According to the probability model, one can cut the edges with small frequencies so the complex TSP becomes simpler. The probability model works well even though for TSP on sparse graphs. We apply the special frequency quadrilaterals to the family of TSP instances constructed by Cornuejols and Nemhauser [28] who verifies the worst performance ratio of the Christofieds algorithm for metric TSP. After certain edges are cut according to their frequencies, the OHC can be found in $O(n)$ time based on the preserved graph.

This paper is organized as follows. First in Sect. 2, we shall introduce the special frequency quadrilaterals. In Sect. 3, a probability model for the OHC edges is built based on special frequency quadrilaterals. In Sect. 4, we apply the probability model to the family of metric TSP constructed by Cornuejols and Nemhauser to trim the useless edges. In Sect. 5, the conclusions are drawn and the possible future research work is given.

2   The Special Frequency K_4s

Before the computation of special frequency quadrilaterals, we first review the general frequency quadrilaterals. Given four vertex $\{A, B, C, D\}$ in K_n, they form a quadrilateral $ABCD$. The $ABCD$ contains six edges (A,B), (A,C), (A,D), (B,C), (B,D), and (C,D), and their distances are $d(A,B)$, $d(A,C), d(A,D), d(B,C), d(B,D),$ and $d(C,D)$, respectively. Under the assumption that the three sum distances $d(A,B) + d(C,D)$, $d(A,C) + d(B,D)$, and $d(A,D) + d(B,C)$ are unequal, Wang and Remmel [25] first derive six optimal four-vertex paths with given endpoints, and then compute a general frequency quadrilateral with the six optimal four-vertex paths. For an edge $e \in \{(A,B), (A,C), (A,D), (B,C), (B,D), (C,D)\}$, the frequency $f(e)$ is the number of optimal four-vertex paths containing e. Since the three sum distances have six permutations, there are six frequency quadrilaterals for quadrilateral $ABCD$. In each frequency $ABCD$, the frequency $f(e)$ of an edge e is either 1, 3, or 5. In addition, the frequency $f(e) = 1$, 3, and 5 related to edge e occur twice, respectively, in the six frequency quadrilaterals. Let $p_i(e)$ denote the probability that e has frequency $i \in \{1, 3, 5\}$ in a frequency quadrilateral containing e. The probability $p_1(e) = p_3(e) = p_5(e) = \frac{1}{3}$ based on the six frequency quadrilaterals. Edge e is contained in $\binom{n-2}{2}$ quadrilaterals in K_n. Given a quadrilateral containing e, the corresponding frequency quadrilateral will be one of the six frequency quadrilaterals. Therefore, the probability that e has the frequency $f(e) = 1$, 3, and 5 in a random frequency quadrilateral is $\frac{1}{3}$, respectively. For an OHC edge $e_o = (A,B)$ in K_n, Wang and Remmel [25] found $n - 3$ quadrilaterals where e_o has the frequency 3 or 5 rather than 1. In the rest frequency quadrilaterals, we assume the conservative probability $p_1(e_o) = p_3(e_o) = p_5(e_o) = \frac{1}{3}$ for e_o.

The probability that e_o has the frequency 1, 3, and 5 in a random frequency quadrilateral is derived as formula (1).

$$p_1(e_o) = \frac{1}{3} - \frac{2}{3(n-1)} \text{ and}$$

$$p_3(e_o) = \frac{1}{3} \text{ and}$$

$$p_5(e_o) = \frac{1}{3} + \frac{2}{3(n-1)}. \tag{1}$$

Given a quadrilateral $ABCD$, we usually meet the cases that the three sum distances $d(A,B)+d(C,D)$, $d(A,C)+d(B,D)$, and $d(A,D)+d(B,C)$ are equal or two of them are equal. In these cases, the six frequency quadrilaterals $ABCD$ in paper [25] will change as well as the probability model for the OHC edges e_o. Firstly, we assume two of the three sum distances are equal. There are six cases:
  (a) $d(A,B)+d(C,D) = d(A,D)+d(B,C) < d(A,C)+d(B,D)$,
  (b) $d(A,B)+d(C,D) = d(A,D)+d(B,C) > d(A,C)+d(B,D)$,
  (c) $d(A,B)+d(C,D) = d(A,C)+d(B,D) < d(A,D)+d(B,C)$,
  (d) $d(A,B)+d(C,D) = d(A,C)+d(B,D) > d(A,D)+d(B,C)$,
  (e) $d(A,C)+d(B,D) = d(A,D)+d(B,C) < d(A,B)+d(C,D)$, and
  (f) $d(A,C)+d(B,D) = d(A,D)+d(B,C) > d(A,B)+d(C,D)$.
It mentions that a frequency quadrilateral $ABCD$ is computed with the six optimal four-vertex paths in $ABCD$, see [25]. As some of the sum distances $d(A,B)+d(C,D)$, $d(A,C)+d(B,D)$, and $d(A,D)+d(B,C)$ are equal, the set of six optimal four-vertex paths is not unique. For example the case (a), there are two sets of six optimal four-vertex paths. (1) If we assume $d(A,B)+d(C,D) < d(A,D)+d(B,C)$, then the six optimal four-vertex paths are (A,B,C,D), (B,C,D,A), (C,D,B,A), (D,A,B,C), (B,A,C,D), and (A,B,D,C). (2) Else if $d(A,B)+d(C,D) > d(A,D)+d(B,C)$, then the six optimal four-vertex paths are (A,B,C,D), (B,C,D,A), (C,D,B,A), (D,A,B,C), (B,C,A,D), and (A,D,B,C). Based on the two sets of optimal four-vertex paths, the two frequency $ABCD$s are computed and illustrated in Fig. 1 (1) and (2), respectively. For the two frequency $ABCD$s (1) and (2) in Fig. 1, we assume they occur with the equal probability $\frac{1}{2}$ for case (a). In this case, the expected frequencies of the six edges are computed as $f(A,B) = f(C,D) = f(A,D) = f(B,C) = 4$ and $f(A,C) = f(B,D) = 1$, respectively. Considering the expected frequency of each edge, the special frequency quadrilateral $ABCD$ for case (a) is represented as Fig. 2(a). For the residual cases (b)-(f), the same method is used to derive the optimal four-vertex paths and compute the expected frequencies of edges. The corresponding special frequency quadrilaterals are computed and illustrated in Fig. 2(b)-(f), respectively. The distance inequalities are given below each of the corresponding special frequency quadrilaterals.

Secondly, as the three sum distances $d(A,B)+d(C,D)$, $d(A,C)+d(B,D)$, and $d(A,D)+d(B,C)$ are equal, one has to order them for computing the six optimal four-vertex paths and the unique frequency quadrilateral $ABCD$. The

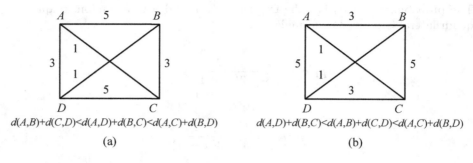

Fig. 1. Two frequency $ABCD$s for case (a)

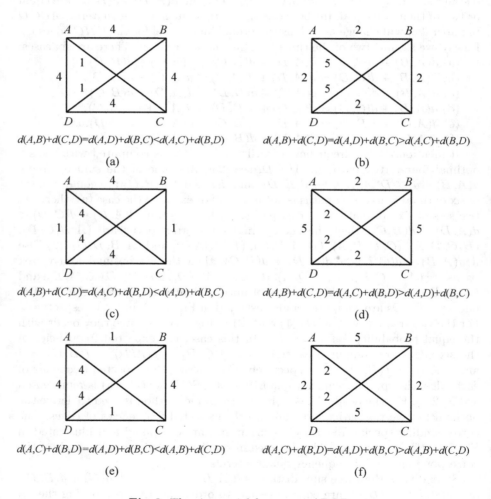

Fig. 2. The six special frequency $ABCD$s

general frequency quadrilaterals [25] or the special frequency quadrilaterals in Fig. 2 can be used for this kind of quadrilaterals.

3   The Probability Model for OHC Edges e_o

Different from the general frequency quadrilaterals in [25], the edges do not always have the frequencies 1, 3, and 5 in each special frequency quadrilateral. For an edge e, such as (A, B), it has the frequency 1, 2, 4, or 5 in the six special frequency quadrilaterals. Moreover, the frequencies 2 and 4 appear 2 times, respectively, and the frequencies 1 and 5 appear once, respectively. Let $p_i(e)$ denote that e has the frequency $i \in \{1, 2, 4, 5\}$ in a special frequency quadrilateral containing e. Obviously, the $p_1(e) = \frac{1}{6}$, $p_2(e) = \frac{1}{3}$, $p_4(e) = \frac{1}{3}$, and $p_5(e) = \frac{1}{6}$ hold based on the six special frequency quadrilaterals. Thus, the average frequency of e over the six special frequency quadrilaterals is 3. As edge e in K_n is contained in $\binom{n-2}{2}$ quadrilaterals $ABCD$ where the two of the three sum distances are equal, it is contained in $\binom{n-2}{2}$ special frequency $ABCD$s as well. Each special frequency $ABCD$ will have the equal probability to be one of the six special frequency quadrilaterals in Fig. 2. Therefore, given a special frequency $ABCD$ containing a common edge e, the probability $p_1(e) = p_5(e) = \frac{1}{6}$ and $p_2(e) = p_4(e) = \frac{1}{3}$.

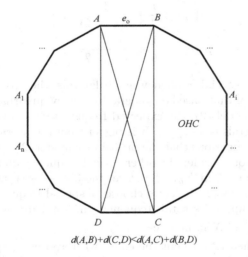

$$d(A,B)+d(C,D)<d(A,C)+d(B,D)$$

Fig. 3. The special $ABCD$s for an OHC edge $e_o = (A, B)$

For an OHC edge e_o, the probability will be different. Figure 3 shows the OHC of K_n and a quadrilateral $ABCD$. The edge $e_o = (A, B)$ and (C, D) belong to the OHC. Since (A, B) and (C, D) belong to the OHC, the sum distance $d(A, B) + d(C, D)$ is less than $d(A, C) + d(B, D)$. Otherwise, edges (A, B) and

(C, D) will be replaced by edges (A, C) and (B, D) in the OHC. Here we consider the special quadrilaterals $ABCD$ where two of the three sum distances are equal. There are two cases $d(A, B) + d(C, D) = d(A, D) + d(B, C)$ and $d(A, C) + d(B, D) = d(A, D) + d(B, C)$. Plus the inequality $d(A, B) + d(C, D) < d(A, C) + d(B, D)$, we derive the other two inequalities $d(A, B) + d(C, D) = d(A, D) + d(B, C) < d(A, C) + d(B, D)$ and $d(A, B) + d(C, D) < d(A, C) + d(B, D) = d(A, D) + d(B, C)$. The two inequalities correspond to the special frequency quadrilaterals (a) and (f) in Fig. 2. In Fig. 2, the frequency of $e_o = (A, B)$ in the two special frequency $ABCD$s is 4 and 5, respectively. Based on the two special frequency $ABCD$s, the probability $p_4(e_o) = p_5(e_o) = \frac{1}{2}$ whereas the other probabilities are equal to zero. Thus, the average frequency of $e_o = (A, B)$ over the two special frequency $ABCD$s is 4.5. There are $n - 3$ nonadjacent edges $(C, D) \in OHC$ for edge $e_o = (A, B)$. In the residual frequency quadrilaterals containing e_o, we assume the probability $p_1(e_o) = p_5(e_o) = \frac{1}{6}$ and $p_2(e_o) = p_4(e_o) = \frac{1}{3}$. Thus, the probabilities $p_1(e_o)$, $p_2(e_o)$, $p_4(e_o)$, and $p_5(e_o)$ for e_o are derived as formula (2).

$$p_1(e_o) = \frac{1}{6} - \frac{1}{3(n-2)} \text{ and }$$

$$p_2(e_o) = \frac{1}{3} - \frac{2}{3(n-2)} \text{ and }$$

$$p_4(e_o) = \frac{1}{3} + \frac{1}{3(n-2)} \text{ and }$$

$$p_5(e_o) = \frac{1}{6} + \frac{2}{3(n-2)}. \tag{2}$$

According to the special frequency quadrilaterals, the probability model (2) for e_o is different from that based on general frequency quadrilaterals [25]. Based on the probability model (2), the expected frequency of e_o over the six special frequency quadrilaterals is $3 + \frac{3}{n-2}$. It is bigger than the average frequency 3 of a common edge e. It mentions that the probability model (2) is very conservative for e_o since we did not consider the other special frequency quadrilaterals where e_o has frequency 4 and 5 in K_n. Therefore, the $3 + \frac{3}{n-3}$ is just a lower bound of the average frequency of e_o. As we choose N special frequency quadrilaterals containing e_o to compute its total frequency, the total frequency of e_o will be bigger than $\left(3 + \frac{3}{n-2}\right) N$ in most cases.

This gives us a heuristic to cut the edges with average frequency below $3 + \frac{3}{n-2}$ and the OHC edges e_o are kept intact. Wang and Remmel [26, 29] have designed the iterative algorithms to reduce a TSP on K_n to the TSP on sparse graphs. In this research, we start from the TSP on sparse graphs where the degrees of vertices are very small. Even in this case, we can still cut the other edges out of OHC based on the probability model (2).

4   An Application of the Probability Model

Christofides [10] first proposes the $\frac{3}{2}$-approximation algorithm that requires polynomial time for the metric TSP. In the next, Cornuejols and Nemhauser [28] constructs a family of metric TSP where the approximation ratio of the Christofides heuristic is exactly $\frac{3}{2}$ in the worst case. Given such a TSP instance, we cut the useless edges according to their average frequencies computed based on the special frequency quadrilaterals. After the useless edges are eliminated, the TSP instance becomes quite simple and the OHC can be found in polynomial time. Since a quadrilateral contains at least four vertices, we assume $n \geq 4$.

Using the rectilinear distances between two vertices, Cornuejols and Nemhauser [28] builds the metric TSP illustrated in Fig. 4, where the distance of an edge (v_i, v_j) is the shortest distance between the two endpoints v_i and v_j. Given three vertices v_i, v_j, and v_k in the graph, the distances satisfy the constrains: (a) $d(v_i, v_j) \geq 0$; (b) $d(v_i, v_j) = d(v_j, v_i)$; (c) $d(v_i, v_j) + d(v_i, v_k) \geq d(v_j, v_k)$. For convenience of analysis, we add two edges (v_1, v_{2m-1}) and (v_3, v_{2m}) in the original graph. The two edges are noted with the thinner lines.

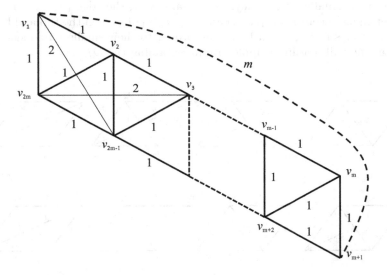

Fig. 4. The graphs defined with rectilinear distances [28]

In the next, we compute the average frequency of each edge based on the special frequency quadrilaterals, and cut the edges with the smallest frequencies for every vertex v_i. In the previous papers [26,29], Wang and Remmel designed the iterative algorithms to cut the useless edges for TSP on K_n. Here, we cut the useless edges for the graph in Fig. 4. It says the probability model (2) works not only for dense graphs of TSP, but also for the TSP sparse graphs. There are two types of vertices according to their degrees. The degree of the first type of vertices is three, such as v_1, v_m, v_{m+1} and v_{2m} in Fig. 4. The degree of the second type of vertices is four, for example, the vertices v_2, v_3, etc. The number of the four-degree vertices is the maximum. For each four-degree vertex, two of the associated edges belong to the OHC, and the other two associated edges are useless.

Without loss of generality, we choose the vertex v_2 to compute the average frequencies of the four associated edges (v_1, v_2), (v_2, v_3), (v_2, v_{2m-1}), and (v_2, v_{2m}). With the help of the two auxiliary edges (v_1, v_{2m-1}) and (v_3, v_{2m}), there are four special quadrilaterals $v_1v_2v_{2m-1}v_{2m}$, $v_1v_2v_3v_{2m}$, $v_1v_2v_3v_{2m-1}$, and $v_2v_3v_{2m-1}v_{2m}$ containing the four edges. Moreover, each of the four edges visiting v_2 is contained in three of the four quadrilaterals. The corresponding special frequency quadrilaterals can be selected from the Fig. 2. The four quadrilaterals and the corresponding special frequency quadrilaterals are shown in Fig. 5. The frequencies of each edge containing v_2 in the four special frequency quadrilaterals are given in Table 1. The average frequency of each edge is also computed and illustrated at the bottom of Table 1. According to the average frequencies of the four edges, the two edges (v_1, v_2) and (v_2, v_3) should be preserved based on the probability model (2). The other two edges (v_2, v_{2m-1}) and (v_2, v_{2m}) can be cut due to the smallest frequency. In the same way, the edge (v_1, v_{m+1}) and all the other edges except for (v_1, v_{2m}) and (v_m, v_{m+1}) between the top and bottom horizontal edges in Fig. 4 can be cut according to their smallest frequencies. In this case, one will obtain a simple graph containing the OHC.

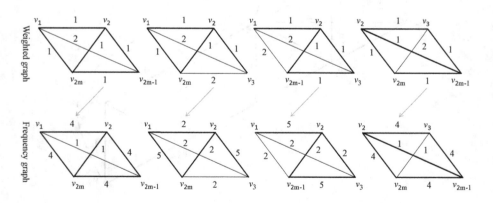

Fig. 5. The four special frequency quadrilaterals containing v_2

In Fig. 3 of paper [28], vertex v_1 is contained in 5 edges. One can build the 10 special frequency quadrilaterals containing v_1, and compute the average frequencies of the edges containing v_1. It finds that the average frequency of the two OHC edges is $\frac{7}{2}$ which is bigger than the average frequency $\frac{8}{3}$ of the other three edges. The OHC edges are found according to the frequency of edges.

Table 1. The frequencies of each edge containing v_2 in the four special frequency quadrilaterals

Frequency quadrilateral	$f(v_1, v_2)$	$f(v_2, v_3)$	$f(v_2, v_{2m-1})$	$f(v_2, v_{2m})$
$v_1 v_2 v_{2m-1} v_{2m}$	4	0	4	1
$v_1 v_2 v_3 v_{2m}$	2	5	0	2
$v_1 v_2 v_3 v_{2m-1}$	5	2	2	0
$v_2 v_3 v_{2m-1} v_{2m}$	0	4	1	4
Avg. frequency	$\frac{11}{3}$	$\frac{11}{3}$	$\frac{7}{3}$	$\frac{7}{3}$

5   Conclusions

The special frequency quadrilaterals are presented for a class of quadrilaterals $ABCD$ in K_n where the two of the three sum distances $d(A, B) + d(C, D)$, $d(A, D) + d(B, C)$, and $d(A, C) + d(B, D)$ are equal. As one chooses N such frequency quadrilaterals containing an edge to compute its total frequency, the probability model demonstrates that the frequency of an OHC edge will be bigger than that of a common edge. We apply the special frequency quadrilaterals to the metric TSP instances constructed by Cornuejols and Nemhauser for cutting the useless edges. The special frequency quadrilaterals works well to eliminate all the useless edges. In the residual graph, the OHC can be found in $O(n)$ time.

References

1. Johnson, D.S., McGeoch, L.A.: The Traveling Salesman Problem and Its Variations. Combinatorial Optimization, 1st edn. Springer Press, London (2004). https://doi.org/10.1007/b101971
2. Karp, R.: On the computational complexity of combinatorial problems. Networks **5**(1), 45–68 (1975)
3. Held, M., Karp, R.: A dynamic programming approach to sequencing problems. J. Soc. Ind. Appl. Math. **10**(1), 196–210 (1962)
4. Bellman, R.: Dynamic programming treatment of the traveling salesman problem. J. ACM **9**(1), 61–63 (1962)
5. Carpaneto, C.G., Dell'Amico, M., Toth, P.: Exact solution of large-scale, asymmetric traveling salesman problems. ACM Trans. Math. Softw. **21**(4), 394–409 (1995)
6. Klerk, E.D., Dobre, C.: A comparison of lower bounds for the symmetric circulant traveling salesman problem. Discrete Appl. Math. **159**(16), 1815–1826 (2011)
7. Levine, M.S.: Finding the right cutting planes for the TSP. J. Exp. Algorithmics **5**, 1–16 (2000)
8. Applegate, D., et al.: Certification of an optimal TSP tour through 85900 cities. Oper. Res. Lett. **37**(1), 11–15 (2009)
9. Thomas, H.C., Charles, E.L., Ronald, L.R., Clifford, S.: Introduction to Algorithm, 2nd edn. China Machine Press, Beijing (2006)
10. Christofides, N.: Worst-case analysis of a new heuristic for the traveling salesman problem. Technical report, DTIC Document (1976)

11. Mömke, T., Svensson, O.: Approximating graphic TSP by matchings. In: Proceedings of the 2011 IEEE 52nd Annual Symposium on Foundations of Computer Science (FOCS 2011), pp. 560–569. IEEE, Palm Springs (2011)
12. Helsgaun, K.: An effective implementation of the Lin-Kernighan traveling salesman heuristic. Eur. J. Oper. Res. **126**(1), 106–130 (2000)
13. Sharir, M., Welzl, E.: On the number of crossing-free matchings, cycles, and partitions. SIAM J. Comput. **36**(3), 695–720 (2006)
14. Heidi, G.: Enumerating all Hamilton cycles and bounding the number of Hamiltonian cycles in 3-regular graphs. Electr. J. Comb. **18**(1), 1–28 (2011)
15. Björklund, A., Husfeldt, T., Kaski, P., Koivisto, M.: The traveling salesman problem in bounded degree graphs. ACM Trans. Algorithms **8**(2), 1–18 (2012)
16. Eppstein, D.: The traveling salesman problem for cubic graphs. J. Graph Algorithms Appl. **11**(1), 61–81 (2007)
17. Liśkiewicz, M., Schuster, M.R.: A new upper bound for the traveling salesman problem in cubic graphs. J. Discrete Algorithms **27**, 1–20 (2014)
18. Aggarwal, N., Garg, N., Gupta, S.: A 4/3-approximation for TSP on cubic 3-edge-connected graphs. http://arxiv.org/abs/1101.5586
19. Boyd, S., Sitters, R., van der Ster, S., Stougie, L.: The traveling salesman problem on cubic and subcubic graphs. Math. Program. **144**(1), 227–245 (2014)
20. Correa, J.R., Larré, O., Soto, J.A.: TSP tours in cubic graphs: beyond 4/3. SIAM J. Discrete Math. **29**(2), 915–939 (2015)
21. Borradaile, G., Demaine, E.D., Tazari, S.: Polynomial-time approximation schemes for subset-connectivity problems in bounded-genus graphs. Algorithmica **68**(2), 287–311 (2014)
22. Gharan, S.O., Saberi, A.: The asymmetric traveling salesman problem on graphs with bounded genus. In: Proceedings of the Twenty-Second Annual ACM-SIAM Symposium on Discrete Algorithms (SODA 2011), pp. 1–12. ACM (2011)
23. Jonker, R., Volgenant, T.: Nonoptimal edges for the symmetric traveling salesman problem. Oper. Res. **32**(4), 837–846 (1984)
24. Hougardy, S., Schroeder, R.T.: Edge elimination in TSP instances. In: Kratsch, D., Todinca, I. (eds.) WG 2014. LNCS, vol. 8747, pp. 275–286. Springer, Cham (2014). https://doi.org/10.1007/978-3-319-12340-0_23
25. Wang, Y., Remmel, J.B.: A binomial distribution model for the traveling salesman problem based on frequency quadrilaterals. J. Graph Algorithms Appl. **20**(2), 411–434 (2016)
26. Wang, Y., Remmel, J.B.: An iterative algorithm to eliminate edges for traveling salesman problem based on a new binomial distribution. Appl. Intell. **48**(11), 4470–4484 (2018)
27. Wang, Y.: An approximate method to compute a sparse graph for traveling salesman problem. Expert Syst. Appl. **42**(12), 5150–5162 (2015)
28. Cornuejols, G., Nemhauser, G.L.: Tight bounds for Christofides' traveling salesman heuristic. Math. Program. **14**(1), 116–121 (1978)
29. Wang, Y., Remmel, J.: A method to compute the sparse graphs for traveling salesman problem based on frequency quadrilaterals. In: Chen, J., Lu, P. (eds.) FAW 2018. LNCS, vol. 10823, pp. 286–299. Springer, Cham (2018). https://doi.org/10.1007/978-3-319-78455-7_22

Data Science and Machine Learning Theory

Sampling to Maintain Approximate Probability Distribution Under Chi-Square Test

Jiaoyun Yang[1,2], Junda Wang[1,3(✉)], Wenjuan Cheng[1,2], and Lian Li[1,2]

[1] National Smart Eldercare International S&T Cooperation Base,
Hefei University of Technology, Hefei, Anhui, China
{jiaoyun,llian}@hfut.edu.cn
[2] School of Computer Science and Information Engineering,
Hefei University of Technology, Hefei, Anhui, China
cheng@ah.edu.cn
[3] School of Mathematics, Hefei University of Technology, Hefei, Anhui, China
wjd980325@126.com

Abstract. In data management center, sometimes it is necessary to provide a subset to show data characteristics, among which probability distribution is an important one. Sampling is a fundamental method to generate data subsets. But how to sample a minimum subset with fixed approximation ratio of probability distributions is still a problem. In this paper, we define the approximation ratio as the significant difference level in chi-square test and use this test to formulate the sampling problem. We decompose the probability distribution as conditional probabilities based on Bayesian networks and propose a heuristic search algorithm to generate the subset by designing two scoring functions, which are based on chi-square test and likelihood functions, respectively. Experiments on four types of datasets with size 60000 show that when setting significant difference level α to 0.05, the algorithm could exclude 99.5%, 97.5%, 84.8% and 90.8% samples based on their Bayesian networks, respectively.

Keywords: Sampling · Chi-square test · Bayesian network · Probability distribution

1 Introduction

Sampling is a fundamental problem in data science, especially in the big data era. It can generate small size subsets to represent the original whole datasets to reduce the computational complexity, and has been widely applied to many applications. An important one is data trading or data exchange [1]. In data center, data sellers or suppliers often need to provide a subset to show the data characteristics. An intuition behind this is that if the distributions of the subset and the whole dataset are consistent or similar, the subset could show most of the characteristics of the whole dataset. Data sellers or suppliers could even

© Springer Nature Singapore Pte Ltd. 2019
X. Sun et al. (Eds.): NCTCS 2019, CCIS 1069, pp. 29–45, 2019.
https://doi.org/10.1007/978-981-15-0105-0_3

provide the subset rather than the original whole dataset to meet data buyers' or demander's analysis requirement. In this paper, we focus on the sampling methods that could guarantee the distributions.

Based on statistical properties, the sampling methods could be divided into two categories, probability sampling and nonprobability sampling [2]. The difference between them is whether some elements have no chance to be selected. The most typical probability sampling is random sampling, in which each element has an equal probability to be chosen. However, its performance is instable. Various sampling methods are designed to improve the sampling quality. Systematic sampling is to sort elements according to some rules and then choose elements at regular intervals [3]. Stratified sampling divides elements into various categories or strata and applies random sampling in each category at specific sampling ratio [4]. The element partition should minimize the varieties within categories and maximize varieties between categories. Clustering methods could be adopted for this partition task [5,6]. But it is different from cluster sampling, which chooses a whole cluster at a time. Generally, these probability sampling methods are all random sampling in their processes. When the size of the chosen subsamples becomes larger, the distribution will become closer to the original distribution. But there are no mechanisms in these methods to minimize the subpopulations' size while maintaining the probability distribution.

For nonprobability sampling, methods are often designed for particular applications. In social science, researchers usually use snowball sampling [7,8] to recruit subjects. An initial subjects group is first determined, and then more subjects are recruited based on previous chosen subjects. In data science area, some researchers try to determine the minimum sampling size within the fixed data analysis performance [9–11]. Silva et al. used the machine learning model performance as the criteria to choose samples by a heuristic search approach [9]. Alwosheel et al. determined the minimum size for artificial neural networks by Monte Carlo experiments [10]. These works are for particular machine learning models. When the model is changed, the sampled subsets may become unsatisfactory with the criteria.

In order to maintain the probability distribution when sampling, we need to conduct hypothesis testing. However, the joint distribution of the original dataset is usually hard to determine. Judea Pearl developed Bayesian networks to represent the relationships between attributes. Hereby, the joint distribution could be decomposed into the product of several conditional probabilities based on Bayesian networks. With the conditional probabilities, researchers could infer posterior probability by Markov Chain Monte Carlo (MCMC) sampling [12–14] or Gibbs sampling [15,16]. These sampling methods could also generate samples that satisfy the original joint distribution. They first initial a random sample and then generate new samples based on the conditional probabilities. These generated samples may not exist in the original dataset, as they are generated by probabilities.

In this paper, we aim to develop sampling methods such that the distribution of the sampled subset is close to the original distribution as well the size of the subset is minimized. We first define α-approximate probability distribution under chi-square test based on Bayesian networks. With this distribution constraint, we propose a heuristic search algorithm to find the minimum size subset by designing scoring functions. The performance is tested on four different types of datasets, which are generated based on four Bayesian networks, including ASIA, ALARM, HEPAR2, and ANDES. Results show that when setting significant difference level α to 0.05, the proposed algorithm could exclude 99.5%, 97.5%, 84.8% and 90.8% samples from original datasets with size 60000, respectively, while the constraints of chi-square test are still satisfied.

The remain of this paper is organized as follows. In Sect. 2, we introduce related works including Bayesian networks and its scoring functions. In Sect. 3, the problem is formulated and the sampling method is introduced. The experiment results are described in Sect. 4, and we summarize the paper in the last section.

2   Related Works

2.1   Bayesian Networks

A Bayesian network is a probabilistic graphical model [17], which organizes the attributes into a directed graph (DAG). In the graph, each node represents a variable or attribute. Assume there are n variables, i.e.$\{x_1, x_2, ..., x_n\}$, then there should be n nodes in the graph, which is denoted as B_n. A directed edge from x_i to x_j means that the values of x_j depend on that of x_i. x_i is usually named as a parent node, and x_j is named as a children node. One node could have multiple parent nodes and multiple children nodes. Here, we use $\pi(x_i)$ to stand for the parent node set of x_i. For each node, the graph defines a conditional probability table based on its parent nodes. Figure 1 shows a Bayesian network example. In the graph, there are 5 nodes. For each node, there is a conditional probability table, which denotes the variable's probability values conditioning on its parent nodes.

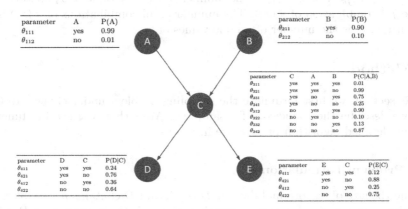

Fig. 1. A Bayesian network example. For each node, there is a conditional probability table, in which $\theta_{ijk} = P(x_i = k | \pi(x_i) = j)$. It denotes the probability of kth value of node x_i conditioning on the jth combination of its parent nodes $\pi(x_i)$.

The conditional probability tables could be regarded as the parameters of a Bayesian network. Here, we use θ_{ijk} to encode the parameters, which denotes the probability of kth value of node i conditioning on the jth combination of its parent nodes. With these tables, the joint probability distribution could be factorized as the product of all conditional probability distributions in the network, which can be written as Eq. 1 [18].

$$P(x_1, \ldots, x_n) = \prod_{i=1}^{n} P(x_i | \pi(x_i)) \tag{1}$$

2.2   Scoring Function

Given a dataset, we often need to evaluate the quality of a constructed Bayesian network based on score functions. These functions could be classified into two categories: Bayesian-based functions and information-theoretic-based functions. For the computational efficiency purpose, these scores are defined on the network structure and parameters, which allows for efficient leaning algorithms based on local search methods. The general idea of Bayesian-based scoring functions is to compute the posterior probability distribution from a prior probability distribution on the possible networks conditioning on dataset D. The best network is the one that maximizes the posterior probability [19]. Information-theoretic-based functions require to choose the network with the minimum encoding length among all the networks. CH function is a widely used one [20], which is defined as Eq. 2.

$$LL(D) = \sum_{i=1}^{n} \sum_{j=1}^{q_i} \frac{\Gamma\left(\sum_{k=1}^{r_i} \theta_{ijk}\right)}{\Gamma\left(M_{ij} + \sum_{k=1}^{r_i} \theta_{ijk}\right)} \sum_{k=1}^{r_i} \frac{\Gamma(M_{ijk} + \theta_{ijk})}{\Gamma(\theta_{ijk})} \tag{2}$$

where D is the original data, and M is the size of D. We can obtain $M_{ijk} = \theta_{ijk} * \sum_{k=1}^{r_i} M_{ijk}$, where M_{ijk} is the number of the kth value of x_i conditioning on the jth combination of $\pi(x_i)$. The number of all combinations of $\pi(x_i)$ is q_i. In addition, r_i is the number of unique values of x_i.

3   Method

In this section, we first formulate the sampling problem under chi-square test, then we describe the framework for sampling. After that, the scoring function and the detailed algorithm process are introduced.

3.1   Problem Formulation

Assume the size of the original dataset D is M, and D contains n variables. Let B be the corresponding Bayesian network of D. θ_{ijk} is the parameter of B, which could be determined based on D by $\theta_{ijk} = P(x_i = k | \pi(x_i) = j)$. It denotes the probability of the kth value of node x_i conditioning on the jth combination of its parent nodes $\pi(x_i)$. We aim to choose a subset D' such that the distributions of D' and D are close enough. Here we use the chi-square test to determine the similarity of these two distributions. So the sampling problem is to find a subset D' with the minimum size m that satisfy the chi-square test, which is denoted as Eqs. 3 and 4.

$$test(D'; D; i, j) = [\sum_{k=1}^{r_i} \frac{\left(m_{ijk} - \theta_{ijk} * \sum_{k=1}^{r_i} m_{ijk}\right)^2}{\theta_{ijk} * \sum_{k=1}^{r_i} m_{ijk}} < \chi_{1-\alpha}(p-1)] \qquad (3)$$

$$test'(D'; D; i) = [\sum_{j=1}^{q_i} \sum_{k=1}^{r_i} \frac{\left(m_{ijk} - \eta_{ijk} * m\right)^2}{\eta_{ijk} * m} < \chi_{1-\alpha}(p-1)] \qquad (4)$$

In these two equations, η_{ijk} is the joint distribution of node x_i and its parent nodes $\pi(x_i)$ in dataset D, which can be calculated according to $P(x_i, \pi(x_i))$. m_{ijk} is the number of the kth value of x_i conditioning on the jth combination of $\pi(x_i)$ in dataset D'. q_i is the number of all possible combinations of $\pi(x_i)$. r_i is the number of unique values of x_i. α is the confidence level, and $1 - \alpha$ denotes the significant difference level. p is the degree of freedom in chi-square distribution.

These two equations are used for conducting chi-square test on node x_i. The joint probability distribution of dataset D could be decomposed into the product of conditional probabilities specified by the corresponding Bayesian network. As long as the conditional probabilities of datasets D and D' are the same, then the joint probability distributions are equal. Therefore, the distribution test could be conducted on the probabilities of the network. The probabilities is defined on each node, so the probability distribution test could only happen on each node. For node x_i, there are conditional probability θ_{ijk} and joint probability η_{ijk}. Both of these two probabilities need to be tested. This is why the problem has two constraints. Equations 3 and 4 are used to test conditional probability θ_{ijk} and joint probability η_{ijk}, respectively. In these two equations, θ_{ijk} and η_{ijk} are calculated based on dataset D, which represent the distribution of D, and m_{ijk} and m are from dataset D', which represent the distribution of D'.

$\chi_{1-\alpha}(p-1)$ is the pre-fixed chi-square test threshold value, which could be achieved with α and p by checking the chi-square distribution table. If the chi-square value is less than this threshold, we could say these two distributions have no significant difference with α confidence level. Both of these two equations have variable p, which denote the degree of freedom. In Eq. 3, p is equal to the number of possible values of x_i. In Eq. 4, p is determined by Eq. 5. Equations 3 and 4 are used to test node x_i's conditional distribution and joint distribution, respectively. Hereby, these two p have different values.

$$p = q_i * r_i - \sum_{j=1}^{q_i} \sum_{k=1}^{r_i} I\left(\theta_{ijk} = 0\right) \qquad (5)$$

When θ_{ijk} is equal to zero, the value of function I is one otherwise zero.

3.2   Sampling Algorithm

The main idea of the proposed sampling algorithm is to apply heuristic search to find a dataset D' with the minimum size m based on the defined scoring functions. First, D' is initialized to empty set, then data samples are added into D' based on scoring functions gradually. The process is listed as follows.

- Step 1: Enumerate all the combinations of $\pi(x_i)$ for each node x_i, and encode these combinations.
- Step 2: Initialize an empty set D', and calculate θ_{ijk} and η_{ijk} based on dataset D and the given Bayesian network structure.
- Step 3: Generate C subsets by randomly sampling, each subset with size $|\Delta D'|$. Evaluate these C subsets by scoring functions, and add the one with the best scoring value into D'.
- Step 4: Sort all samples in D' based on scoring functions, and delete the worst β samples.
- Step 5: If the distributions of all the nodes in the Bayesian network satisfy the problem constraints, i.e. Eqs. 3 and 4, then terminate. Otherwise, goto step 3.

In step 1, we enumerate all the combinations of $\pi(x_i)$ for each node x_i and encode them. These codes are used as j in θ_{ijk} and η_{ijk}. Step 2 initializes D' and calculates θ_{ijk} and η_{ijk}, which represent the distributions of D and will be used for chi-square test. In step 3, we randomly sampling C subsets of D and evaluate them based on the scoring functions, which will be described in the next subsection. From these C subsets, we add the best one into D'. We choose to add a subset with size $|\Delta D'|$ rather than a sample into D' in order to accelerate the sampling process. After that, step 4 checks the D' according to scoring functions and eliminates some samples with poor scoring values. As the subsets are randomly sampled, some samples in the subset may not be good enough. Therefore, step 4 will try to exclude these samples from D'. The samples are added into D' gradually. At first, D' may not satisfy the constraints. With the size of D' growing, it will meet the requirement at some point. Step 5 are used to determine whether D' satisfies the requirement. If so, then terminate the process, otherwise goto step 3 to continue to add samples into D'.

3.3   Scoring Function

In this section, we introduce how to evaluate a sample or a sample set by defining scoring functions. Here, two scoring functions are used, i.e. Eqs. 6 and 7, which are derived from chi-square test and Dirichlet distribution assumption, respectively. These two scoring functions are used to evaluate the similarity of the distributions of D and D'. Equation 6 applies $test(D'; D; i, j))$ and $test'(D'; D; i)$ to check whether the jth conditional probability distribution of x_i and the joint probability distribution of x_i and $\pi(x_i)$ satisfy the chi-square test, respectively. The calculations are based on Eqs. 3 and 4. If test passes, the value of $I()$ function equals to 0, otherwise 1. Therefore, Eq. 6 could find how many nodes could not pass the chi-square test.

$$L(D, D') = \sum_{i=1}^{n} (\sum_{j=1}^{r_i} (I(test(D'; D; i, j))) + I(test'(D'; D; i)) \tag{6}$$

$$W(D, D') = \sum_{i=1}^{n} \sum_{j=1}^{q_i} \sum_{k=1}^{r_i} M_{ijk} \log \left(\frac{m_{ijk}}{\sum_{k=1}^{r_i} m_{ijk}} \right) \tag{7}$$

In Eq. 7, M_{ijk} and m_{ijk} are the number of the kth value of x_i conditioning on the jth combination of $\pi(x_i)$ in dataset D and D', respectively. This equation assumes samples in the dataset obey the Dirichlet distribution. It denotes the likelihood value given the distribution parameter γ' of D', i.e. $W(D, D') = \log(P(D|D') = \log(P(D|\gamma')$. Let us determine the value of γ'.

γ' is the Dirichlet distribution parameter of D', hereby, for a sample d_l in D', its probability is defined as Eq. 8.

$$P(d_l|\gamma) = \prod_{i=1}^{n} \prod_{j=1}^{q_i} \prod_{k=1}^{r_i} \chi(i, j, k; d_l)^{\gamma_{ijk}} \tag{8}$$

where

$$\chi(i, j, k; d_l) = \begin{cases} 1, & x_i = k, \pi(x_i) = j \\ 0, & otherwise \end{cases} \tag{9}$$

Then the likelihood function of d_l is Eq. 10.

$$\log(P(d_l|\gamma)) = \sum_{i=1}^{n} \sum_{j=1}^{q_i} \sum_{k=1}^{r_i} \chi(i, j, k; d_l) \log(\gamma_{ijk}) \tag{10}$$

The likelihood function of D' is Eq. 11.

$$\log(P(D'|\gamma)) = \sum_{i=1}^{n} \sum_{j=1}^{q_i} \sum_{k=1}^{r_i} m_{ijk} \log(\gamma_{ijk}) \tag{11}$$

Based on Gibb's inequality, if $P(x)$ and $Q(x)$ are two probability distributions over the same domain, then $\sum_x P(x) \log(Q(x)) \leq \sum_x P(x) \log(P(x))$. So in order to maximize the likelihood function Eq. 11, γ'_{ijk} must satisfy the following equation.

$$\gamma'_{ijk} = \frac{m_{ijk}}{\sum_{k=1}^{r_i} m_{ijk}} \tag{12}$$

Therefore,

$$W(D, D') = \log(P(D|\gamma') = \sum_{i=1}^{n} \sum_{j=1}^{q_i} \sum_{k=1}^{r_i} M_{ijk} \log(\gamma'_{ijk})$$

$$= \sum_{i=1}^{n} \sum_{j=1}^{q_i} \sum_{k=1}^{r_i} M_{ijk} \log \left(\frac{m_{ijk}}{\sum_{k=1}^{r_i} m_{ijk}} \right)$$

According to Gibb's inequality, $W(D, D')$ achieves the maximum value only if γ'_{ijk} satisfies Eq. 13, which means the distributions of D and D' are the same.

$$\gamma'_{ijk} = \theta_{ijk} = \frac{M_{ijk}}{\sum_{k=1}^{r_i} M_{ijk}} \tag{13}$$

We add a subset $\Delta D'$ into D' in each iteration. The evaluation should be conducted for $\Delta D'$. Equations 14 and 15 denote this evaluation. Equation 15 is defined as $W(D, D', \Delta D') = \log(\mathrm{P}(D|D' \cup \Delta D')) - \log(\mathrm{P}(D|D'))$. Based on Eq. 7, we could obtain Eq. 15. m^*_{ijk} is the number of the kth value of x_i conditioning on the jth combination of $\pi(x_i)$ in dataset $\Delta D'$.

$$L(D, D', \Delta D') = \sum_{i=1}^{n} (\sum_{j=1}^{r_i} (I(test(D' \cup \Delta D'; D; i, j))) + I(test'(D' \cup \Delta D'; D; i)) \tag{14}$$

$$W(D, D', \Delta D') = \sum_{i=1}^{n} \sum_{j=1}^{q_i} \sum_{k=1}^{r_i} M_{ijk} * \log(\frac{m_{ijk} + m^*_{ijk}}{m_{ijk}} * \frac{\sum_{k=1}^{r_i} m_{ijk}}{\sum_{k=1}^{r_i} (m_{ijk} + m^*_{ijk})}) \tag{15}$$

Equation 14 denotes how many nodes do not pass the chi-square test after adding $\Delta D'$ into D', while Eq. 15 reflects the increase in scoring values after the adding operation. Therefore, we should find $\Delta D'$ with smaller value of $L(D, D', \Delta D')$ based on Eq. 14 and choose $\Delta D'$ with larger value of $W(D, D', \Delta D')$ based on Eq. 15. When selecting $\Delta D'$, we first compare the values of $L(D, D', \Delta D')$. If the values of $L(D, D', \Delta D')$ are the same, we then compare the values of $W(D, D', \Delta D')$. In addition, in order to accelerate the sampling process, we only count for nodes that do not pass the chi-square test in the calculation of Eq. 15.

3.4   Encoding

In the sampling algorithm, we need to calculate all the values of m_{ijk} for scoring functions. The value of m_{ijk} will change if we add a sample. So, we need to compute these values repetitively during each iteration, which is quite time-consuming. Actually, we only need to calculate the change of m_{ijk}, which will reduce the time complexity. Here, we apply an encoding strategy to fulfill this task.

In m_{ijk}, i denotes the ith variable, j is the combination serial number of x_i's parent nodes $\pi(x_i)$, and k is the value of x_i. For a given sample d, i and k could be easily determined. The most difficult part is determining j. Assume x_i has $|\pi(x_i)|$ parent nodes, i.e. $\pi(x_i) = \{x_i^1, x_i^2, ..., x_i^{|\pi(x_i)|}\}$. Assume there are r_i^l possible values for parent node x_i^l. In sample d, the values of $\pi(x_i)$ are $\{k_i^1, k_i^2, ..., k_i^{|\pi(x_i)|}\}$. Then the combination serial number j could be determined by Eq. 16.

$$j = ((((k_i^1 * r_i^1) + k_i^2) * r_i^2 + \cdots + k_i^{|\pi(x_i)|-1}) * r_i^{|\pi(x_i)|-1} + k_i^{|\pi(x_i)|} \tag{16}$$

For each sample, the combination serial number of parent nodes for each node x_i could be calculated during the pre-processing. This value will not change during the algorithm. So we could store these values in a array. When updating m_{ijk}, we could only check this array to accomplish the task.

3.5   Time Complexity Analysis

In this section, we first list the pseudocodes, then analyze the time complexity. Algorithm 1 shows the main process of the sampling algorithm.

Algorithm 1. Sampling Algorithm

Input: dataset D, Bayesian network B, C, β
Output: subset D'
1 $D' = \varnothing$;
2 **m=0**;
3 (indexJ, θ, η)=**preProcess**(D, B);
4 **for** $h = 0; h < M/(|\Delta D'| - \beta); h + +$ **do**
5 $\quad$ bestChiScore=∞; bestDScore=$-\infty$; bestM=**0**;
6 $\quad$ **for** $hh = 0; hh < C; hh + +$ **do**
7 $\quad\quad$ $\Delta D'$=randomSampling(D);
8 $\quad\quad$ **m***=**distCal**($\Delta D'$,indexJ);
9 $\quad\quad$ (chiScore, dScore)=**scoreCal**(**m, m***, θ, η);
10 $\quad\quad$ **if** *(chiScore<bestChiScore or (chiScore=bestChiScore and dScore>bestDScore))* **then**
11 $\quad\quad\quad$ bestChiScore=chiScore;
12 $\quad\quad\quad$ bestDScore=dScore;
13 $\quad\quad\quad$ bestM=**m***;
14 $\quad\quad\quad$ *best$\Delta D'$=$\Delta D'$*;
15 $\quad$ **m=m+bestM**;
16 $\quad$ $D' = D' \cup best\Delta D'$; $D = D - best\Delta D'$;
17 $\quad$ $D' = D' - worstSample(D', \beta)$; $D = D \cup worstSample(D', \beta)$;
18 $\quad$ **m=m-distCal**(worstSample(D', β));
19 $\quad$ **if** *bestChiScore=0* **then**
20 $\quad\quad$ **break**;

In the input of Algorithm 1, C denotes the randomly generated subset number in each iteration. β represents the size of samples that need to be excluded from D' after each iteration. Line 1 initializes D' as an empty set. Line 2 initializes an array **m** with 0, which is used to record the distribution of D'. Line 3 calls procedure *preProcess* to determine the distribution of D and the encoding value of each node x_i's parent nodes for each sample. The distribution is recorded by θ and η, and the encoding values are recorded by *indexJ*. The loop of line 4–20 is the main sampling process. In each iteration, the algorithm extracts at least $|\Delta D'| - \beta$ samples from D, so there are at most $M/(|\Delta D'| - \beta)$ iterations. The loop of line 6–14 generates C subsets and evaluates them based on scoring functions. Line 7 extracts $|\Delta D'|$ samples randomly from D, then the distribution of $\Delta D'$ is calculated in line 8 by calling procedure $distCal$. Line 9 calls procedure $scoreCal$ to compute the chi-square test scoring value and likelihood scoring value. Line 10 judges whether current subset $\Delta D'$ is better by comparing these two scoring values. If so, line 11–14 record the scoring values and the distribution of this subset. Line 15–16 are used to add $\Delta D'$ into D' by joining the datasets and combining the distributions. Line 17–18 delete the worst β samples from D' and add them back into D. We indicate in Sect. 3.3 that the scoring function 14 represents how many nodes do not satisfy the chi-square test. Therefore, when this function value equals to 0, the subset D' meets the requirement. Line 19 makes this judgment and terminates the algorithm.

Algorithm 2. preProcess

Input: dataset D, Bayesian network B
Output: indexJ, θ, η

1   **for** $h = 0; h < |D|; h + +$ **do**
2   **for** $i = 0; i < n; i + +$ **do**
3   **for** $hh = 1; hh <= |\pi(x_i)| - 1; hh + +$ **do**
4        $indexJ(h, i) = (indexJ(h, i) + k_i^{hh}) * r_i^{hh};$
5        $totalCom(i) = (totalCom + r_i^{hh} - 1) * r_i^{hh};$
6       $indexJ(h, i) = indexJ(h, i) + k_i^{|\pi(x_i)|};$
7       $totalCom(i) = totalCom + r_i^{|\pi(x_i)|} - 1;$

8   **for** $h = 0; h < |D|; h + +$ **do**
9   **for** $i = 0; i < n; i + +$ **do**
10      $\theta_{i, indexJ(h,i), d_{h,i}} + +;$

11   **for** $i = 0; i < n; i + +$ **do**
12   **for** $j = 0; j <= totalCom(i); j + +$ **do**
13   sum=0;
14   **for** $k = 0; k < r_i; k + +$ **do**
15   sum=sum+$\theta_{ijk};$
16   **for** $k = 0; k < r_i; k + +$ **do**
17       $\eta_{ijk} = \theta_{ijk}/|D|;$
18       $\theta_{ijk} = \theta_{ijk}/sum;$

Algorithm 2 describes the detailed procedure of pre-processing. The main purpose of pre-processing is to calculate the distribution parameters θ and η of D, as well as the encoding value of the combinations of x_i'parent nodes for each sample. As in the main process of the sampling algorithm, m_{ijk} needs to be computed repetitively during each iteration, including line 8 and 18 in Algorithm 1. So, we calculate the encoding value j before the main loop to reduce the time complexity. In Algorithm 2, line 1–7 accomplish this task based on Eq. 16. In line 1, h indicates the sample index number, which ranges from 0 to $|D|$. In line 2, i represents the variable index number, which ranges from 0 to n. In line 3, hh is used to mark x_i's hh^{th} parent node. k_i^{hh} and r_i^{hh} denote the values and ranges of this x_i's hh^{th} parent node, respectively. $totalCom(i)$ records the total combination number of the values of x_i's parent nodes. $d(h, i)$ is the value of x_i in the hth sample. line 8–18 calculate θ and η for dataset D. In Algorithm 2, Line 4, 5, 6 and 7 can be accomplished in $O(1)$ time. Assume the largest value of $totalCom(i)$ and r_i are q and r, respectively. $\pi(x_i)$ in line 3 should be smaller than $totalCom(i)$. So Line 1–7 needs $O(|D|*n*q)$ time. Line 8–9 needs $O(|D|*n)$ time. Line 11–18 need $O(n*q*r)$ time. $preProcess$ procedure needs $O(|D|*n*q+|D|*n+n*q*r) = O(|D|*n*q+n*q*r)$ time.

Algorithm 3. distCal

> **Input:** dataset $\Delta D'$, indexJ
> **Output:** m*
> 1 **for** $ii = 0; ii < |\Delta D'|; ii + +$ **do**
> 2 $\quad$ **for** $i = 0; i < n; i + +$ **do**
> 3 $\quad\quad$ h=index(ii);
> 4 $\quad\quad$ $m^*_{i,indexJ(h,i),d(h,i)} = m^*_{i,indexJ(h,i),d(h,i)} + 1;$

Algorithm 3 calculates **m*** for dataset $\Delta D'$, which represents the distribution and will be used in the calculation of score values. m^*_{ijk} is the number of the kth value of node x_i conditioning on the jth combination of its parent nodes $\pi(x_i)$ in $\Delta D'$. So the main loop of this procedure is to check each sample in $\Delta D'$ and update m^*_{ijk} by finding the corresponding value. Line 3 finds the index number in D for sample ii of $\Delta D'$ as the encoding value j is indexed based on the index numbers of samples in D. With this number, we could find the encoding value j of the values of x_i's parent nodes for sample ii. $d(h, i)$ is x_i's value in sample ii. In this procedure, line 3–4 need $O(1)$ time. So the whole procedure needs $O(|\Delta D'| * n)$ time.

Algorithm 4 presents the detailed procedure of the calculation of scoring functions based on Eqs. 14 and 15. The inputs **m, m***, θ, and η denote the distribution of D', $\Delta D'$, and D, respectively. Line 1 conducts the summation calculation to obtain the distribution of the joint set of D' and $\Delta D'$. The loop of line 2–16 calculates the chi-square test scoring value. This is a nested loop, where i ranges from 0 to n, j ranges from 0 to q_i, k ranges from 0 to r_i. n, q_i, and r_i denote the number of variables, the total combination number of x_i's parent nodes, and the number of unique values of x_i, respectively. There are two chi-square test in the problem formulation, i.e. Eqs. 3 and 4. $chiSquare1$ records the first test, and $chiSquare2$ records the second test. The tests are conducted on each variable x_i. If a variable passes the tests, then chi-square test score value equals to 0. In order to accelerate the sampling process, the likelihood score values only count those variables that do not pass the chi-square tests. These variables are stored in array $node$ in Algorithm 4. Line 15–16 and line 19–20 conduct this operation. After that, line 21–29 compute the likelihood score value on these variables. In order to be in consistent with Algorithm 2, we assume the maximum value of

q_i and r_i are q and r, respectively. The arithmetic operations in Algorithm 4 needs $O(1)$ time. *node* only stores variables and do not need to maintain the ordered relationship, therefore, line 18 and 19 also needs $O(1)$ time. Hereby, line 5–16 need $O(r)$ time, and line 2–20 need $O(n*q*r)$ time. There are at most n variables in *node*, so line 21–29 need $O(n*q*r)$ time. The *scoreCal* procedure needs $O(n*q*r)$ time in all.

Algorithm 4. scoreCal

Input: m, m*, θ, η, indexJ
Output: chiScore, dScore

1 $\hat{m} = m + m^*$;
2 **for** $i = 0; i < n; i++$ **do**
3   chiSquare2=0;
4   **for** $j = 0; j < q_i; j++$ **do**
5          $sum_m = sum_{m^*} = 0$;
6   chiSquare1=0;
7   **for** $k = 0; k < r_i; k++$ **do**
8              $sum_m = sum_m + m_{ijk}$; $sum_{m^*} = sum_{m^*} + m^*_{ijk}$;
9          $sum = sum_m + sum_{m^*}$; $size = |D'| + |\Delta D'|$;
10   **for** $k = 0; k < r_i, k++$ **do**
11             $chiSquare1 += (\hat{m}_{ijk} - \theta_{ijk} * sum)^2/(\theta_{ijk} * sum)$;
12             $chiSquare2 += (\hat{m}_{ijk} - \eta_{ijk} * size)^2/(\eta_{ijk} * size)$;
13   **if** $chiSquare1 >= \chi_{1-\alpha}(p_1 - 1)$ **then**
14   chiScore+=1;
15   **if** $vis(i) = 0$ **then**
16   node.pushback(i);   vis(i)=1;
17   **if** $chiSquare2 >= \chi_{1-\alpha}(p_2 - 1)$ **then**
18   chiScore+=1;
19   **if** $vis(i) = 0$ **then**
20   node.pushback(i); vis(i)=1;
21 **for** $ii = 0; ii < node.size; ii++$ **do**
22   i=node(ii);
23   **for** $j = 0; j < q_i; j++$ **do**
24         $sum_m = sum_{m^*} = 0$;
25   **for** $k = 0; k < r_i; k++$ **do**
26             $sum_m = sum_m + m_{ijk}$; $sum_{m^*} = sum_{m^*} + m^*_{ijk}$;
27         $sum = sum_m + sum_{m^*}$;
28   **for** $k = 0; k < r_i; k++$ **do**
29             $dScore += \theta_{ijk} * |D| * \log(\hat{m}_{ijk} * sum_m/(m_{ijk} * sum))$;

Now we can analyze the whole time complexity. Line 3 calls *preProcess* procedure, which needs $O(|D| * n * q + n * q * r)$ time. Line 7 needs $O(|\Delta D'|)$ to generate a subset. Line 8–9 call *distCal* and *scoreCal* procedures, which needs $O(|\Delta D'|*n)$ and $O(n*q*r)$, respectively. So line 7–14 need $O(|\Delta D'|+|\Delta D'|*n+n*q*r) = O(|\Delta D'|*n+n*q*r)$ time. Line 6–14 need $O(C*(|\Delta D'|*n+n*q*r))$ time. **m** is an array with size $n*q*r$, so line 15 needs $O(n*q*r)$. Line 16 needs $O(|\Delta D'|)$ and line 17 needs $O(\beta)$. Line 18 needs $O(n*q*r + \beta*n)$. C, $|\Delta D'|$, β are pre-fixed constant values. So line 5–20 need $O(n*q*r)$ time. The whole sampling algorithm needs $O(|D|*n*q+n*q*r+|D|*n*q*r) = O(|D|*n*q*r)$ time.

4   Experiment

In the experiment, we want to test how many samples are needed to meet the requirement of distribution tests. We decompose the original distribution as the product of conditional probabilities based on Bayesian networks. Hereby, Bayesian network's structure should be determined first. Here we use four Bayesian networks for our experiments, including ASIA, ALARM, HEPAR2 and MUNI1, which are described below.

1. **ASIA:** The ASIA network [21] is a small network, which is related to chest disease diagnosis, including tuberculosis, lung cancer, or bronchitis. This network contains 8 nodes and 8 edges. Each node corresponds to a binary variable.
2. **ALARM:** The ALARM network [22] is a medium network, which is also related to disease diagnosis. This network consists of 37 nodes and 46 edges. The number of unique values for each variable could be 2, 3 or 4.
3. **HEPAR2:** The HEPAR2 network [23] is a large network, which is used for the diagnosis of liver disorders. It consists of 70 nodes and 123 edges.
4. **ANDES:** ANDES [24] ANDES is a very large network, which is used in an intelligent tutoring system. It consists of 223 nodes and 338 edges.

For each network, we first applied Gibbs sampler to generate 3 datasets with different sizes, including 20000, 40000 and 60000. In the sampling algorithm, $|\Delta D'|$ is set as 120, i.e. we add 120 samples into D' in each iteration. Also, we delete at most 20 samples from D' at the end of each iteration, which means $\beta = 20$.

Three metrics are applied to evaluate the algorithm, including sampled size, running time (in second) and averaged distribution difference. Sampled size represents the size of subset that are extracted from the original dataset to meet the chi-square test requirement. Averaged distribution difference is calculated based on Eq. 17, where $num(\theta_{ijk})$ is the total number of θ_{ijk}. This value reflects the distribution difference between the sampled subset and the original dataset.

$$delta = \frac{1}{num(\theta_{ijk})} \sum_{i=1}^{n} \sum_{j=1}^{q_i} \sum_{k=1}^{r_i} |\frac{m_{ijk}}{\sum_{k=1}^{r_i} m_{ijk}} - \theta_{ijk}| \tag{17}$$

We first set the significant difference level α as 0.05. Table 1 summarizes the sampling results in terms of sampled size, running time and averaged distribution difference. We can find that the sampling algorithm could exclude most of the samples from original datasets, but still maintain the distribution. For the datasets with size 60000, we could exclude 99.5%, 97.5%, 84.8% and 90.8% samples from original datasets for these four networks, respectively. When the Bayesian network structure becomes more complex, the sampled subset also becomes larger. This is because when the structure becomes more complex, the distribution of the original dataset also becomes more complex. More data are needed to construct the complex distribution. Similarly, when the size of the original dataset becomes larger, small probabilities have more chances to happen, which makes the distribution become more complex. This means it also needs more samples to satisfy the requirement. For AISA network, the sampled sizes of datasets with size 40000 and 60000 are equal. This is because this network is small and the distribution is simple. It is very easy to satisfy Eqs. 3 and 4. Besides, we can find that all the averaged distribution difference are around 0.05, which is the fixed significant difference level. This validates the effectiveness of our sampling algorithm.

Table 1. Experiment results on 3 different size datasets based on 4 Bayesian networks in terms of sampled size, running time, and averaged distribution difference, when $\alpha = 0.05$

	20000			40000			60000		
	Size	Time	Delta	Size	Time	Delta	Size	Time	Delta
ASIA	100	7	0.004	300	22	0.004	300	25	0.004
ALARM	1200	75	0.036	1400	102	0.041	1500	259	0.046
HEPAR2	4100	259	0.062	6100	632	0.067	9100	1401	0.071
ANDES	2200	312	0.032	3200	564	0.031	5500	1101	0.027

size, time, and delta denote the size of sampled subset, running time, and averaged difference in distribution, respectively

Table 2 shows the results on the same datasets as Table 1 when setting $\alpha = 0.01$. We can draw similar conclusions as Table 1, i.e. the sampled size becomes larger when the network structure becomes more complex or the size of the original dataset becomes larger. The sampled size is generally smaller than that in Table 1 for the same dataset. This is because the significant difference level α is lower in Table 2, which allows more differences in the distribution.

As the samples in subset $\Delta D'$ are randomly selected, some samples may have poor scoring function values. This is why we delete β worst samples from D' in each iteration. If a sample d_l has good scoring function value, then its rank in D' should be stable, which means when adding new samples into D', these samples' scores should be worse than that of d_l, and d_l's rank should not change a lot. However, if a sample has poor scores, its rank in D' may become higher when adding new samples. We choose 6 samples in the ALARM dataset with size 40000 to illustrate this phenomenon. Figure 2 shows this result. The scoring values are calculated based on Eq. 15. Samples with serial number 1, 7, 130 have stable ranks, while samples with serial number 11, 14, 73 have unstable ranks. The latter samples are those that we should exclude from the sampled subset.

Table 2. Experiment results on 3 different size datasets based on 4 Bayesian networks in terms of sampled size, running time, and averaged distribution difference, when $\alpha = 0.01$

	20000			40000			60000		
	Size	Time	Delta	Size	Time	Delta	Size	Time	Delta
ASIA	100	6	0.021	100	12	0.041	100	16	0.052
ALARM	1200	65	0.051	1200	71	0.053	1400	256	0.051
HEPAR2	3700	215	0.081	5900	616	0.085	8700	1341	0.080
ANDES	1600	127	0.032	2500	330	0.042	3700	708	0.06

*size, time, and delta denote the size of sampled subset, running time, and averaged difference in distribution, respectively

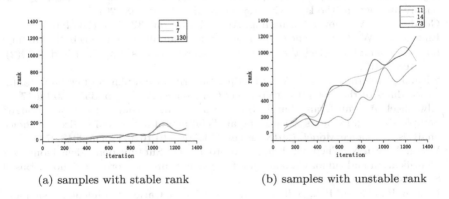

(a) samples with stable rank   (b) samples with unstable rank

Fig. 2. Ranking of different samples in sampled datasets based on scoring functions

5   Conclusion

In this paper, we formulate a sampling problem that requires to find a minimum subset while maintaining the probability distribution under chi-square test. The intuition behind this is that if a subset has similar distribution as the original dataset, it could reflect most of the characteristics of the original dataset, which is usually required in data management center. By decomposing the distribution into conditional probabilities based on Bayesian networks, we proposed a heuristic-search-based sampling algorithm to solve this problem. Two scoring functions are defined based on chi-square test and likelihood functions. Experiments on four different types of datasets show the effectiveness of our sampling algorithm.

Acknowledgment. This work was supported partially by the National Key R&D Program of China (No. 2018YFB1003204), Anhui Provincial Key Technologies R&D Program (No. 1804b06020378, No. 1704e1002221). And the National "111 Project" (No. B14025).

References

1. Goodhart, C.A.E., O'Hara, M.: High frequency data in financialmarkets: issues and applications. Empir. Financ. **4**(2–3), 73–114 (1997)
2. Lohr, S.L.: Sampling: Design and Analysis. Nelson Education, Toronto (2009)
3. Yates, F.: Systematic sampling. Philos. Trans. R. Soc. Lond. Ser. A Math. Phys. Sci. **241**(834), 345–377 (1948)
4. Neyman, J.: On the two different aspects of the representative method: the method of stratified sampling and the method of purposive selection. J. R. Stat. Soc. **97**(4), 558–625 (1934)
5. Rand, W.M.: Objective criteria for the evaluation of clustering methods. J. Am. Stat. Assoc. **66**(336), 846–850 (1971)
6. Aljalbout, E., Golkov, V., Siddiqui, Y., et al.: Clustering with deep learning: taxonomy and new methods. arXiv preprint arXiv:1801.07648 (2018)
7. Goodman, L.A.: Snowball sampling. Ann. Math. Stat. **32**, 148–170 (1961)
8. Emerson, R.W.: Convenience sampling, random sampling, and snowball sampling: how does sampling affect the validity of research? J. Vis. Impair. Blind. **109**(2), 164–168 (2015)
9. Silva, J., Ribeiro, B., Sung, A.H.: Finding the critical sampling of big datasets. In: Proceedings of the Computing Frontiers Conference. ACM, pp. 355–360 (2017)
10. Alwosheel, A., van Cranenburgh, S., Chorus, C.G.: Is your dataset big enough? Sample size requirements when using artificial neural networks for discrete choice analysis. J. Choice Model. (2018). https://doi.org/10.1016/j.jocm.2018.07.002
11. Wang, A., An, N., Chen, G., Liu, L., Alterovitz, G.: Subtype dependent biomarker identification and tumor classification from gene expression profiles. Knowl.-Based Syst. **146**, 104–117 (2018)
12. Paxton, P., Curran, P.J., Bollen, K.A., et al.: Monte Carlo experiments: design and implementation. Struct. Equ. Model. **8**(2), 287–312 (2001)
13. Gilks, W.R., Richardson, S., Spiegelhalter, D.: Markov Chain Monte Carlo in Practice. Chapman and Hall/CRC, London (1995)
14. Wu, S., Angelikopoulos, P., Papadimitriou, C., et al.: Bayesian annealed sequential importance sampling: an unbiased version of transitional Markov chain Monte Carlo. ASCE-ASME J. Risk Uncertain. Eng. Syst. Part B: Mech. Eng. **4**(1), 011008 (2018)
15. George, E.I., McCulloch, R.E.: Variable selection via Gibbs sampling. J. Am. Stat. Assoc. **88**(423), 881–889 (1993)
16. Martino, L., Read, J., Luengo, D.: Independent doubly adaptive rejection metropolis sampling within Gibbs sampling. IEEE Trans. Sig. Process. **63**(12), 3123–3138 (2015)

17. Murphy, K.: An introduction to graphical models. Technical report, University of California, Berkeley, May 2001
18. Friedman, N., Geiger, D., Goldszmidt, M.: Bayesian network classifiers. Mach. Learn. **29**(2–3), 131–163 (1997)
19. Carvalho, A.M.: IST, TULisbon/INESC-ID. Scoring functions for learning Bayesian networks INESC-ID Technical report 54, April 2009
20. Cooper, G.F., Herskovits, E.: A Bayesian method for the induction of probabilistic networks from data. Mach. Learn. **9**(4), 309–347 (1992)
21. Lauritzen, S., Spiegelhalter, D.: Local computations with probabilities on graphical structures and their application on expert systems. J. R. Stat. Soc. **50**, 157–224 (1988)
22. Beinlich, I., Suermondt, G., Chavez, R., Cooper, G.: The ALARM monitoring system: a case study with two probabilistic inference techniques for belief networks. In: Proceedings of the Second European Conference on Artificial Intelligence in Medicine (1989)
23. Onisko, A., Druzdzel, M.J., Wasyluk, H.: A probabilistic causal model for diagnosis of liver disorders. In: Proceedings of the Seventh International Symposium on Intelligent Information Systems (IIS 1998), p. 379 (1998)
24. Conati, C., Gertner, A.S., VanLehn, K., Druzdzel, M.J.: On-line student modeling for coached problem solving using Bayesian networks. In: Jameson, A., Paris, C., Tasso, C. (eds.) User Modeling. ICMS, vol. 383, pp. 231–242. Springer, Vienna (1997). https://doi.org/10.1007/978-3-7091-2670-7_24

Adjusting the Inheritance of Topic for Dynamic Document Clustering

Ruizhang Huang[1,2]([✉]), Yingxue Zhu[1,2], Yanping Chen[1,2], Yue Yang[1,2],
Weijia Xu[1,2], Jian Yang[1,2], and Yaru Meng[1]

[1] College of Computer Science and Technology, Guizhou University, Guiyang,
Guizhou, China
rzhuang@gzu.edu.cn, zhuyingxue1993@gmail.com, ypench@gmail.com,
finfinte007_yyang@163.com, 1316721141@qq.com, 18884985031@163.com,
engyarubella@163.com
[2] Guizhou Provincial Key Laboratory of Public Big Data, Guizhou University,
Guiyang, Guizhou, China

Abstract. Organizing streaming documents from time-varying dataset
is meaningful but difficult because topics evolve over time. Dynamic document clustering is a vital research problem, which helps to group the
time-varying documents into a number of clusters corresponding to their
underlying topics. Datasets are partitioned into a set of time slides to
transfer the streaming document clustering from a continuous problem
to a categorical one. Traditional dynamic document clustering approach
tends to inherit topic information over time directly with no consideration of the nature of datasets. In this paper, we design a novel prior-
adjusted dynamic document clustering approach, namely PADC, which
is able to adjust the topic inheritance process according to two important
of datasets characteristics, in particular, the interval between dataset
time slides and the size of dataset time slides. A collapsed Gibbs sampling algorithm is investigated to infer the document structure for all
time slides with underlying time-varying topics. Parameters for underlying topics inheritance, as well as parameters of the number of clusters
in each time slide, are estimated simultaneously. Extensive experiments
have been conducted comparing the PADC model with state-of-the-art
dynamic document clustering approaches. Experimental results demonstrate that the PADC model is robust and effective for the dynamic
document clustering problem.

Keywords: Topic model · Dynamic clustering · Text mining ·
Topic inheritance

1 Introduction

Nowadays, we are surrounded by overwhelming quantities of textual documents
which can be collected in a streaming manner, e.g. short texts from twitter, standard articles from news websites, etc [11]. Organizing streaming text documents

© Springer Nature Singapore Pte Ltd. 2019
X. Sun et al. (Eds.): NCTCS 2019, CCIS 1069, pp. 46–63, 2019.
https://doi.org/10.1007/978-981-15-0105-0_4

in terms of topics is useful and is heavily needed by many applications. For example, in the application of hotspot analysis, news documents accumulated daily are needed to be grouped to into hotspot topics. However, discovering topics from these large amount of text data is difficult because topics evolve with time. Documents are not exchangeable given their topics but are related with evolving topic content. For example, the news article "Hilary becomes presidential candidate" is with the same topic "2016 U.S. president election"of article "Hilary enters the first election television debate". However, the content of the news topic changes dramatically. New topical words, such as "television" and "debate" were emphasized because the topic is developing over time. Therefore, it is helpful to develop a text document clustering method to discovery document structures with time-varying topics.

Dynamic document clustering, which captures the structure of time-varying document datasets, has received more and more attention due to its research significance and great application potential [5, 7, 15, 23, 30]. Most of existing dynamic clustering approaches are designed based on probabilistic topic models [1, 6, 14]. Datasets are partitioned into a set of time slides to transfer the streaming document clustering from a continuous problem to a categorical one. Topics are sequentially tied through time slides. For each time slide, topics are inherited by taking the topical parameters of last time slide as the prior. Topics are jointly learnt from the topical prior and text documents in each time slide. However, Traditional dynamic document clustering models are designed to learn time-evolving topics with no consideration of the nature of datasets. In reality, the level of topic inheritance is affected greatly in two situations. The first situation is that datasets are partition into time slides with long time intervals. The longer the time interval, the less important the prior. The second situation is that each time slide contains a large number of documents (or words), when the size of each time slide is large, topical prior count less to the current topic estimation. For example, in hotspot analysis, a news event topic usually last for about two weeks. When the time interval is longer than a week, topic content will dramatically change and should be learnt mostly from the current datasets. Therefore, it is necessary to adjust the inheritance of topic for dynamic document clustering by considering the interval between time slides and the size of time slides.

In this paper, we organize a novel prior-adjusted dynamic document clustering approach namely, PADC, to partition documents into time-evolving underlying topics. Two factor are designed to adjust the topical inheritance performance with the consideration of intervals between time slides and the size of time slides. The probability of new topic emerging is also adjusted according to two important of datasets characteristics, in particular, the interval between dataset time slides and the size of dataset time slides. A Gibbs sampling algorithm is developed to learn topical parameters. The number of clusters is learnt automatically with the Gibbs sampling algorithm. We have conducted extensive experiments on our proposed approach by using both synthetic and realistic data sets. We also compared our approach with state-of-the-art models. Experimental results show that our proposed PADC model is robust and effective for dynamic document clustering with topic inheritance.

The remainder of this paper is organized as follows: Sect. 2 reviews the related work on the dynamic topic model. In Sect. 3, we present PADC model and the Gibbs sampling algorithm. Section 4 describes the design of experiments and presents the experimental results. We finally present the conclusion and future work in Sect. 5.

2   Related Work

In the past years, topic modeling has attracted many research efforts from both academia and industry. A topic is a cluster of words frequently co-occurred. Probabilistic topic models, such as probabilistic latent semantic analysis (PLSA) [9], latent dirichlet allocation (LDA) [3], Hierarchical Dirichlet Process (HDP) [20], and Latent dirichlet mixture model (LDMM) [4], are based upon the idea that documents are mixtures of topics, where a topic is a probability distribution over words. The PLSA model [9] is designed with the assumption that each document contains different topics and each word is generated by one of these topics. The LDA model used a dirichlet prior to solve the over-fitting problem which makes the model more flexible [3]. In LDMM model [4], the topic probabilities are represented by a mixture of dirichlet distributions.

Topic models have played vital role in discovering latent semantic structures from complex data corpora, ranging for text documents to web news articles [10,19,21,25,27]. With the spreading of social media, extensions to LDA have been proposed to identify topics of blogs [10,19,31], twitters [17,21], and analyze sentiment orientation of reviews [26,27]. For example, a biterm topic model (BTM) is able to effectively model short texts and mitigate the sparse-feature problem in short texts [28]. The joint sentiment topic model (JST) is proposed to detect sentiment and topic simultaneously from reviews [8]. Twitter-LDA model [32] analyzes topics on Twitter.

To capture dynamic topic from text streams (such as newspaper and twitter), a number of dynamic topic models have been proposed to consider topic evolution over time [2,16,21,24]. The dynamic topic model (DTM) [2] designed based on LDA, is investigated to analyze to documents with time-varying topics. The dynamic mixture model (DMM) [24] considers a single dynamic sequence of documents, which corresponds to a single topic over time. The topic over time model (ToT) [22] needs all samples over time for the inference process. It cannot be updated sequentially, and is not appropriate for data that are continuously accumulated. The topic tracking model (TTM) [13] focuses on tracking time-varying consumer behavior, in which consumer interests change over time [31]. The dynamic clustering topic model (DCT) [16] extends a collapsed Gibbs Sampling algorithm for the dirichlet multinomial mixture model (GSDMM) [29] to time evolution scenario, which derives a good influence on short text dynamic clustering. Global and local topic model (GLTM) [17] treats different types of topics differently. It detects global and local topics simultaneously by using geo-tag Twitter data.

However, no existing work studies research how topics are transformed and inherited in time slides. None of them organize document into time-evolving topics, where topics are inherited from lats time slide. The degree of topical inheritance is adjusted considering the nature of datasets.

3   Prior-Adjusted Dynamic Document Clustering Model

In this section, we propose prior-adjusted dynamic document clustering model, namely PADC, to learn document clustering structure with underlying time-evolving topics. Formally we defined the following terms:

- A word w is an item from a vocabulary indexed by $\{1, 2, ..., V\}$.
- A cluster is characterized by a multinomial distribution over words. It is represented by a multinomial parameter.
- A document d_{t_m} at time t is represented as a V-dimensional vector $d_{t_m} = \{n_{w_1}, n_{w_2}, ..., n_{w_V}\}$ where n_{w_j} is the number of appearance of the word w_j of the document d_{t_m}, and the d_{t_m} represent the m-th document of $\mathbf{D}_t$.
- A document data set $\mathbf{D}_t$ is a collection of M documents at time t denoted by $\mathbf{D}_t = \{d_{t_1}, d_{t_2}, ..., d_{t_M}\}$.

The level of inheritance of the topical proor is adjusted with regard of the interval of time slides and the size of time slides. The probability of emerging new topics is also adjusted accordingly. The graphical representation of our proposed PADC model is the same with the DCT approach [16] which is depicted in Fig. 1. Two adjusting parameters, namely η and γ, are introduced to adjust the topical prior information, where η is the time slide interval adjustment factor, and γ is the time slide size adjustment factor. We analysis the nature of text document datasets in four situations with the consideration of the interval and size of time slides. The shorter the time slide interval and the smaller the time slide size, the stronger the topical information is inherited. The converse is also true. Precisely, when the interval of time slide is longer than a predefined parameter ι, and the size of time slide is larger than a predefined parameter δ, the topical prior. $\alpha_{t,z}$ of topic z at time t is adjusted as follows:

$$\alpha_{t,z} = \begin{cases} \alpha_{t,z} * (1 + \eta + \gamma), & if\ z \in \Omega_{t-1} & (1) \\ \alpha_{t,z}, & if\ z \notin \Omega_{t-1} & (2) \end{cases}$$

where Ω_{t-1} denote the topic set learnt in time $t - 1$.

When the interval of time slide is shorter than ι and the size of time slide is larger than δ, the topical prior, $\alpha_{t,z}$ of topic z at time t is adjusted as follows:

$$\alpha_{t,z} = \begin{cases} \alpha_{t,z} * (1 - \eta + \gamma), & if\ z \in \Omega_{t-1} & (3) \\ \alpha_{t,z}, & if\ z \notin \Omega_{t-1} & (4) \end{cases}$$

When the interval of timeslide is longer than ι and the size of time slide is smaller than δ, the topical prior, $\alpha_{t,z}$ of topic z at time t is adjusted as follows:

$$\alpha_{t,z} = \begin{cases} \alpha_{t,z} * (1 + \eta - \gamma), & if\ z \in \Omega_{t-1} & (5) \\ \alpha_{t,z}, & if\ z \notin \Omega_{t-1} & (6) \end{cases}$$

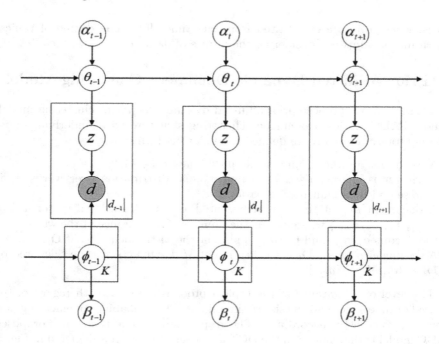

Fig. 1. Graphical representation of PADC model.

When the interval of times lide is shorter than ι and the size of time slide is smaller than δ, the topical prior, $\alpha_{t,z}$ of topic z at time t is adjusted as follows:

$$\alpha_{t,z} = \begin{cases} \alpha_{t,z} * (1 - \eta - \gamma), & if \ z \in \Omega_{t-1} \qquad (7) \\ \alpha_{t,z}, & if \ z \notin \Omega_{t-1} \qquad (8) \end{cases}$$

The generative process of PADC is as follows:

1. Draw $\Theta_t \sim Dirichlet(\alpha_t \Theta_{t-1})$.
2. Fore each topic z Draw $\phi_{t,z} | \beta_{t,z} \phi_{t-1,z} \sim Dirichlet(\beta_{t,z} \phi_{t-1,z})$.
3. For each document d:
   (a) Draw $z_d \sim multinomial(\Theta_t)$
   (b) For each word w_d:
       i. Draw $w_d | \phi_{t,z_d} \sim multinomial(\phi_{t,z_d})$.

where $\alpha_t = \{\alpha_{t,z}\}_{z=1}^K$ is adjusted as shown in Eqs. (1) to (8) by the settings of ι and the δ, $\Theta_{t-1} = \{\theta_{t-1,z}\}_{z=1}^K$ is the topic distribution at time t.

Approach Inference. We employ a collapsed Gibbs sampler [18] to learn the parameters of our proposed PADC model. The joint probability of the current document dataset $\mathbf{d_t}$ is estimated as follows:

$$P(\mathbf{d}_t, \mathbf{z}_t | \Phi_{t-1}, \Theta_{t-1}, \alpha_t, \beta_t) = P(\mathbf{d}_t | \mathbf{z}_t, \Phi_{t-1}, \beta_t) P(\mathbf{z}_t | \Theta_{t-1}, \alpha_t, \gamma, \eta) \qquad (9)$$

Algorithm 1. collapsed Gibbs Sampling

Input: Previous topic distribution Θ_{t-1}
Previous word distribution specific to topics Φ_{t-1}
The time slide interval adjustment factor η
The time slide size adjustment factor γ
A set of short documents $\mathbf{d}_t$ at time t
A Initialized α_t and β_t
A Number of iterations I
Output: Current topic distribution Θ_t
Current word distribution specific to topics Φ_t
Documents and its cluster at time t

1 Initialize m_{t,z_d}, n_{t,z_d}, $n_{t,z_d,w}$ as zero in $\mathbf{d}_t$
2 **for** *iteration = 1 to I* **do**
3   **for** *each document $d \in \mathbf{d}_t$* **do**
4   record the current cluster of d: $z = z_d$
5          $m_{t,z_d} \leftarrow m_{t,z_d} - 1$ and $n_{t,z_d} \leftarrow n_{t,z_d} - N_d$
6   **for** *each word $w \in d$* **do**
7              $n_{t,z_d,w} \leftarrow n_{t,z_d,w} - N_{d,w}$
8   **end**
9   draw z_d from equation (14)
10   update m_{t,z_d} and $n_{t,z_d,w}$
11   **end**
12 **end**
13 compute the parameters Θ_t and Φ_t

---

where Θ_{t-1} is the topic distribution at time $t-1$, Φ_{t-1} is the word distribution over topics at time $t-1$, $\mathbf{d}_t$ is the document set at time t, z_t is the set of topics at time t, the α_t and β_t are the prior (or hyper parameters) at time t.

With the chain rule, the above formula can be changed as follows:

$$P(z_d|\mathbf{z}_{t,-d}, \mathbf{d}_t\Phi_{t-1}, \Theta_{t-1}, \gamma, \alpha_t, \beta_t) = \frac{P(\mathbf{z}_t, \mathbf{d}_t|\Phi_{t-1}, \Theta_{t-1}, \gamma, \eta, \alpha_t, \beta_t)}{P(\mathbf{z}_{t,-d}, \mathbf{d}_t|\Phi_{t-1}, \Theta_{t-1}, \gamma, \eta, \alpha_t, \beta_t)} \propto$$
$$\frac{P(\mathbf{z}_t, \mathbf{d}_t|\Phi_{t-1}, \Theta_{t-1}, \gamma, \eta, \alpha_t, \beta_t)}{P(\mathbf{z}_{t,-d}, \mathbf{d}_{t,-d}|\Phi_{t-1}, \Theta_{t-1}, \gamma, \eta, \alpha_t, \beta_t)} \quad (10)$$

Starting molecular moiety derived from the Eq. (10) can be drawn:

$$P(\mathbf{z}_t, \mathbf{d}_t|\Phi_{t-1}, \Theta_{t-1}, \gamma, \eta, \alpha_t, \beta_t) = P(\mathbf{d}_t|\mathbf{z}_t, \Phi_{t-1}, \beta_t)P(\mathbf{z}_t|\Theta_{t-1}, \gamma, \eta, \alpha_t)$$
$$= \int P(\mathbf{d}_t|\mathbf{z}_t, \Phi_t)P(\Phi_t|\Phi_{t-1}, \beta_t)d\Phi_t \times \int P(\mathbf{z}_t|\Theta_t)P(\Theta_t|\Theta_{t-1}, \gamma, , \eta, \alpha_t)d\Theta_t$$
$$= \int \prod_{d=1}^{|\mathbf{d}_t|} \prod_{i=1}^{N_d} P(w_i, d_i|\phi_{t,z,d_i}) \prod_{z=1}^{K} P(\phi_{t,z}|\phi_{t-1,z}, \beta_t)d\Phi_t \times$$
$$\int \prod_{d=1}^{|\mathbf{d}_t|} P(z_{t,d}|\theta_t)P(\theta_t|\theta_{t-1}, \gamma, \eta, \alpha_t)d\Theta_t \quad (11)$$

From the model generation process, it is known that $P(\Theta_t|\Theta_{t-1}, \gamma, \eta, \alpha_t)$ and $P(\Phi_t|\Phi_{t-1}, \beta_t)$ are the Dirichlet distribution, and $P(\mathbf{z}_t|\Theta_t)$ and $P(\mathbf{d}_t|\mathbf{z}_t, \Phi_t)$ are the Multinomial distribution, it can be derived from the formula (11) as follows:

$$\prod_{z=1}^{K} \frac{\Gamma(\sum_{w=1}^{V} \beta_{t,z,w}\phi_{t-1,z,w})}{\prod_{w=1}^{V} \Gamma(\beta_{t,z,w}\phi_{t-1,z,w})} \times \prod_{z=1}^{K} \frac{\prod_{w=1}^{V} \Gamma(n_{t,z,w}+\beta_{t,z,w}\phi_{t-1,z,w})}{\Gamma(\sum_{w=1}^{V} n_{t,z,w}+\beta_{t,z,w}\phi_{t-1,z,w})} \times$$
$$\frac{\Gamma(\sum_{z=1}^{K} \alpha_{t,z}(1\pm\gamma\pm\eta)\theta_{t-1,z})}{\prod_{z=1}^{K} \Gamma(\alpha_{t,z}(1\pm\gamma\pm\eta)\theta_{t-1,z})} \times \frac{\prod_{z=1}^{K} \Gamma(m_{t,z}+\alpha_{t,z}(1\pm\gamma\pm\eta)\theta_{t-1,z})}{\Gamma(\sum_{z=1}^{K} m_{t,z}+\alpha_{t,z}(1\pm\gamma\pm\eta)\theta_{t-1,z})} \tag{12}$$

Bring formula (12) into (10):

$$\frac{P(\mathbf{z}_t,\mathbf{d}_t|\Phi_{t-1},\Theta_{t-1},\gamma,\eta,\alpha_t,\beta_t)}{P(\mathbf{z}_{t,-d},\mathbf{d}_{t,-d}|\Phi_{t-1},\Theta_{t-1},\gamma,\eta,\alpha_t,\beta_t)} = \frac{\Gamma(m_{t,z}+\alpha_{t,z}(1\pm\gamma\pm\eta)\theta_{t-1,z})}{\Gamma(m_{t,z}+\alpha_{t,z}(1\pm\gamma\pm\eta)\theta_{t-1,z}-1)} \times$$
$$\frac{\Gamma(\sum_{z=1}^{K} m_{t,z}+\alpha_{t,z}(1\pm\gamma\pm\eta)\theta_{t-1,z}-1)}{\Gamma(\sum_{z=1}^{K} m_{t,z}+\alpha_{t,z}(1\pm\gamma\pm\eta)\theta_{t-1,z})} \times \frac{\prod_{w\in d} \Gamma(n_{t,z,w}+\beta_{t,z,w}\phi_{t-1,z,w})}{\prod_{w\in d} \Gamma(n_{t,z,w,-d}+\beta_{t,z,w}\phi_{t-1,z,w})} \times \tag{13}$$
$$\frac{\Gamma(n_{t,z,-d}+\sum_{w=1}^{V} \beta_{t,z,w}\phi_{t-1,z,w})}{\Gamma(n_{t,z}+\sum_{w=1}^{V} \beta_{t,z,w}\phi_{t-1,z,w})}$$

The Γ function has this property: $\Gamma(x) = (x-1)\Gamma(x-1)$, and $\Gamma(x+m) = \prod_{i=1}^{m} (x+i-1)\Gamma(x)$, we estimate the conditional probability of topic $z_{t,d}$ for document d at time t as follows:

$$P(z_{t,d}|\mathbf{z}_{t,-d},\mathbf{d}_t,\Phi_{t-1},\Theta_{t-1},\alpha_t,\beta_t) \propto \frac{m_{t,z}+\alpha_{t,z}(1\pm\gamma\pm\eta)\theta_{t-1,z}-1}{\sum_{k=1}^{K} (m_{t,z}+\alpha_{t,z}(1\pm\gamma\pm\eta)\theta_{t-1,z})-1}$$
$$\times \frac{\prod_{w\in d} \prod_{j=1}^{N_{d,w}} (n_{t,z,w-d}+\beta_{t,z,w}\phi_{t-1,z,w}+j-1)}{\prod_{i=1}^{N_d} (n_{t,z,-d}+i-1+\sum_{w=1}^{V} \beta_{t,z,w}\phi_{t-1,z,w})} \tag{14}$$

where $m_{t,z}$ is the total number of documents in $\mathbf{d}_t$ assigned to topic z, $N_{d,w}$ is the number of word w in the document d, $n_{t,z,w,-d}$ is the total number of the word w assigned to topic z except that in d, and $n_{t,z,-d}$ is the total number of documents assigned to z except d. Note that topical adjust parameters γ and η is determined by time slide interval and size of the dataset, the specific adjustments are shown in Eqs. (1) to (8). Detail explanation of our proposed collapsed Gibbs sampling algorithm for our proposed PADC model is shown in Algorithm 1.

4 Experiments

We study the performance of our proposed approach by two sets of experiments. For the first set of experiments, synthetic datasets are used. For the second set of experiments, our proposed PADC approach is evaluated via real document datasets. Each dataset is divided into three unit time in this paper.

4.1 Evaluation Metric

The normalized mutual information (NMI) is used to evaluate the quality of a clustering solution. NMI is a measure that allows us to make the trade-off between the quality of the clustering and the number of clusters. It is an entropy-based metric that explicitly measures the amount of statistical information shared by the variables representing the output clusters and the ground truth clusters of users. In general, NMI is estimated as follows [33]:

$$NMI = \frac{\sum_{h,l} d_{hl} \log(\frac{D \cdot d_{hl}}{d_h c_l})}{\sqrt{(\sum_h d_h \log(\frac{d_h}{D}))(\sum_l c_l \log(\frac{c_l}{D}))}} \qquad (15)$$

where D is the number of documents, d_h is the number of documents in class h, c_l is the number of documents in cluster l, and d_{hl} is the number of documents in class h as well as in cluster l. The NMI value is 1 when a clustering solution perfectly matches the user-labeled class assignments and close to 0 for a random document partitioning.

4.2 Experiments on Synthetic Datasets

Datasets. We derived 4 synthetic datasets to evaluate the effectiveness of our proposed approach. The set of synthetic datasets is developed to simulate the following 4 scenarios, respectively: (1) dataset with long time slide interval and small time slide size; (2) dataset with short time slide interval and small time slide size; (3) dataset with long time slide interval and small time large size; (4) dataset with short time slide interval and large time slide size. Each synthetic dataset consists of M data points organized in 3 time slides with K underling topics. Each data point contain 30 features generated from feature vocabulary with size V. K multinomial distributions were used to represent latent clusters of features in each dataset. Parameters of the K multinomial distributions, denoted as $\{\pi_1, \cdots, \pi_K\}$, were generated by the stick-breaking approach of dirichlet distribution [20]. In particular, the generative of a data point x_{i^t}, where $i \in \{1, \cdots, V\}$ at time t is as follows:

i. Randomly select a cluster $\pi_{k^t} \in \{\pi_{1^t}, \cdots, \pi_{K^t}\}$;

ii. Draw x_{i^t} $Multinomial(\pi_{k^t}; 30)$.

For datasets with short or long time slide interval, the number of clusters K is set to 5 or 10 respectively. We set $V = 5000$ and $M = 1200$ for datasets with large time slide size, and $V = 1000$ and $M = 600$ for datasets with small time slide size respectively. Details of each dataset are shown in Tables 1 and 2.

Table 1. Details of synthetic datasets with small time slide size, where each topic contains 50 document points. Each time slide of dataset is represented by the cluster index.

Cluster (size)		
Time	*dataset1*	*dataset2*
$t = 1$	1(50), 2(50), 3(50)	1(50), 2(50), 3(50)
$t = 2$	3(50), 4(50), 5(50), 6(50)	1(50), 2(50), 3(50), 4(50)
$t = 3$	6(50), 7(50), 8(50), 9(50), 10(50)	1(50), 2(50), 3(50), 4(50), 5(50)

Table 2. Synthetic data sets with large time slide size, where each topic contains 100 document points. Each time slide of dataset is represented by the cluster index.

Cluster (size)		
Time	*dataset3*	*dataset4*
$t = 1$	1(100), 2(100), 3(100)	1(100), 2(100), 3(100)
$t = 2$	3(100), 4(100), 5(100), 6(100)	1(100), 2(100), 3(100), 4(100)
$t = 3$	6(100), 7(100), 8(100), 9(100), 10(100)	1(100), 2(100), 3(100), 4(100), 5(100)

Settings and Results. We conducted experiments on 4 synthetic datasets to verify the performance of our proposed PADC model. For the comparitive study, experimental performances of the DCT model are also investigated. All experimental settings for PADC model and the DCT model are the same. In particular, we set $K = 30$, $\alpha = 0.5$, $\beta = 0.5$, $\eta = 0.15$ and $\gamma = 0.55$. Each approach was run 10 times. The performance is computed by taking the average of these 10 experiments. Each time, we conduct 2000 iterations in our experiments.

For *dataset1*, the prior is correspond to Eqs. (5) to (6) in our proposed PADC model, the NMIs in the data with long time slide interval and small time slide size acquired by PADC model are **0.991 0.982 0.985**, and the NMIs in this dataset acquired by DCT are **0.925 0.913 0.911**. Figures 2 and 3 depicts estimated labels of data points of this dataset, where blue represents the data points assigned to existing historical topic labels, and brown represents the data points assigned to new topic labels. As shown in Fig. 2, our propose PADC model achieves almost perfect data partition results for all time slides. The number of clusters are estimated correctly. However, the DCT model assigns data point to much number of clusters in Fig. 3. The reason is that topics are transformed from previous time slides with no consideration to the fact of long time slide interval and small time slide size.

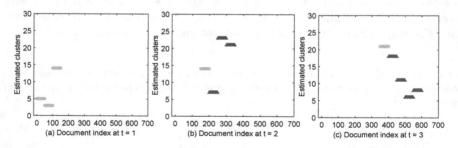

Fig. 2. Estimated labels of the *dataset1* acquired by our proposed PADC model. (Color figure online)

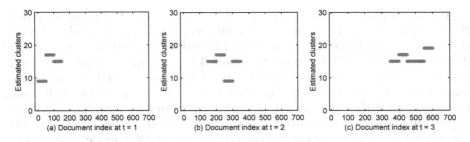

Fig. 3. Estimated labels of the *dataset*1 acquired by DCT model. (Color figure online)

For *dataset*2, the prior is correspond to Eqs. (7) to (8) for our proposed PADC approach, the NMIs in the data with short time slide interval and small time slide size by PADC model are **0.997 0.982 0.988**, and the NMIs in this data set acquired by DCT are **0.912 0.915 0.917**. Figures 4 and 5 depicts estimated labels of data points of this data set, where blue represents the data points assigned to existing historical topic labels, and brown represents the data points assigned to new topic labels. As shown in Fig. 4, our propose PADC model achieves almost perfect data partition results for all time slides. The number of clusters are estimated correctly. However, the DCT model assigns data point to much number of clusters in Fig. 5. The reason is that topics are transformed from previous time slides with no consideration to the fact of short time slide interval and small time slide size.

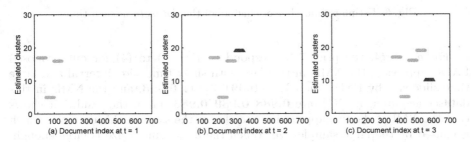

Fig. 4. Estimated labels of the *dataset*2 acquired by our proposed PADC model. (Color figure online)

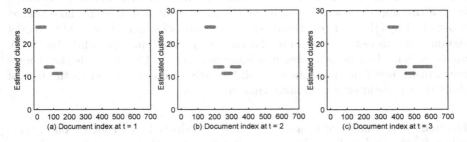

Fig. 5. Estimated labels of the *dataset*2 acquired by DCT model. (Color figure online)

For *dataset*3, the prior is correspond to Eqs. (1) to (2) for our proposed PADC approach, the NMIs in the data with long time slide interval and large time slide size by PADC model are **0.991 0.990 0.996**, and the NMIs in this dataset acquired by DCT are **0.991 0.980 0.971**. From the results, it shows that the NMI of the two approaches are almost the same, which is caused by the large time slide size. The estimated cluster numbers are shown in Fig. 6, from Fig. 6(a), we see that the number of clusters estimated by the two approaches is the same, however, as shown in Fig. 6(b) and (c), our propose PADC model can estimate the number of clusters correctly, while the DCT model assigns data point to much number of clusters. The reason is that topics are transformed from previous time slides with no consideration to the fact of long time slide interval and large time slide size.

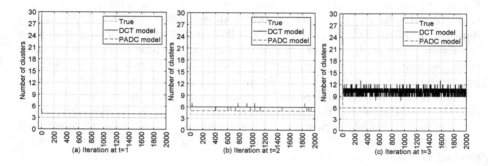

Fig. 6. Estimated number of clusters of the *dataset*3 in each time.

For *dataset*4, the prior is correspond to Eqs. (3) to (4) for our proposed PADC approach, the NMIs in the data with short time slide interval and large time slide size by PADC model are **0.991 0.994 0.996**, and the NMIs in this dataset acquired by DCT are **0.988 0.989 0.983**. From the results, it shows that the NMI of the two approaches are almost the same in this data set, which is caused by the large sample size. When the data sample size is large enough, the weight of topic generation probability will bias on sample, and the impact of historical topic inherited by prior is very weak. The estimated cluster numbers are shown in Fig. 7, from Fig. 7(a), we see that the number of clusters estimated by the two approaches is the same as the number of real clusters, however, as shown in Fig. 7(b) and (c), the number of clusters estimated by PADC model significantly almost approximate the number of real clusters, while the DCT model assigns data point to much number of clusters. The reason is that topics are transformed from previous time slides with no consideration to the fact of short time slide interval and large time slide size.

Discussions. Choice of η and γ: We have conducted experiments on various values for the choice of η and γ in a small field-out synthetic datasets, we observe

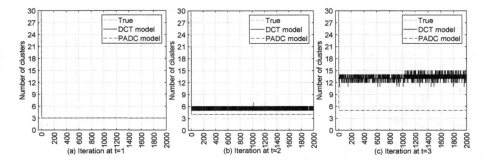

Fig. 7. Estimated number of clusters of the *dataset*4 in each time.

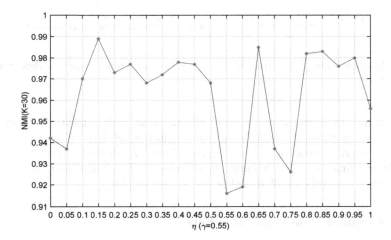

Fig. 8. Average NMI over 3 time slices of the *dataset*2 when η gets different values and $\gamma = 0.55$.

the change of the clustering NMI with another parameter by fixing one of the parameters, We set the value of η to be between 0.1 and 1, with 0.05 as the growth, and the rest of the parameters are set unchanged, and we average the NMI values in the three time slices. Similarly, set the value of γ to be between 0.1 and 1, with 0.05 as the growth. Due to the length of the article, we only listed one experimental result as shown in Figs. 8 and 9. When we set $\eta = 0.15$ and $\gamma = 0.55$, the performance of PADC model is the best compared with other settings. Therefore, we use these experimental settings foe the rest of experiments.

Choice of δ and ι: Parameters ι is used to Help determine the adjustment strategy of the interval of time slide parameter η, which is determined by the interval of time slide, $\eta = 1$ means that the interval of time slide is "day", $\eta = 7$ is "week", $\eta = 30$ is "month", $\eta = 365$ is "year" and so on. And the size of

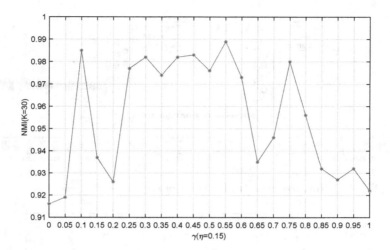

Fig. 9. Average NMI over 3 time slices of the *dataset*2 when γ gets different values and $\eta = 0.15$.

dataset time slide parameter δ, we set the value according to the proportion of the sample size currently time being reached in the total sample size.

Choice of K: Theoretically, we should choose K to be the number of data point. In the process of experiment, we discovered that it is time-consuming. So we choose a relatively small K follow the advice of [12].

4.3   Experiments on Real Data Sets

Data Sets. We conducted experiments on 4 real document datasets for evaluating PADC model,in particular, $Twitter-strong, Twitter-weak, News-strong, News-weak$. The summary of these four text document data sets is shown in Tables 3 and 4. The $Twitter-strong$ and $Twitter-weak$ data sets were derived from the a publicly available corpus of Twitter posts[1] of 2016, where the interval of time slide of $Twitter-strong$ is shorter than ι, and the size of time slide(number of words) is smaller than δ; the interval of time slide of $Twitter-weak$ is longer than ι, and the size(number of words) is smaller than δ. The $News-strong$ and $News-weak$ were derived from the 20$Newsgroups$ collection[2], where the interval of time slide of $News-strong$ is shorter than ι, and the size of time slide (number of words) is larger than δ; the interval of time slide of $News-weak$ is longer than ι, and the size (number of words) is larger than δ.

We preprocessed the $Twitter\text{-}strong, Twitter\text{-}weak, News\text{-}strong, News\text{-}weak$ data sets by stopword removal. High-frequency and Low-frequency words were removed.

---

[1]   https://archive.org/details/twitterstream.
[2]   The description can be found at http://people.csail.mit.edu/jrennie/20Newsgroups.

Table 3. Real data sets with small time slide size.

Time	Twitter-weak		Twitter-strong	
	Cluster	D	Cluster	D
$t = 1$	1, 2, 3	800	1, 2, 3	800
$t = 2$	3, 4, 5, 6	800	1, 2, 3, 4	800
$t = 3$	6, 7, 8, 9, 10	800	1, 2, 3, 4, 5	800

Table 4. Real data sets with large time slide size.

Time	News-weak		News-strong	
	Cluster	D	Cluster	D
$t = 1$	1, 2, 3	800	1, 2, 3	800
$t = 2$	3, 4, 5, 6	800	1, 2, 3, 4	800
$t = 3$	6, 7, 8, 9, 10	800	1, 2, 3, 4, 5	800

Settings and Results. For all set of experiments, we use the same parameter settings of α, β, η and γ of the synthetic data sets to real data sets experiments. For the number of initial clusters K, we set $K = 30$ for all real data sets, and $\iota = 7$, $\delta = 5000$ for our Experimental setup. We compare our model with the following baselines and stat-of-the-art clustering algorithms in our experiments on real-world data sets:

- **dynamic topic model (DTM).** This model utilizes a Gaussian distribution for inferring topic distribution of long text documents in streams.
- **dynamic mixture models (DMM).** This model considers a single dynamic sequence of documents, which corresponds to a single topic over time.
- **topic tracking model (TTM).** Clusters documents based on a dynamic topic tracking model that captures temporal dependencies between long text streams.
- **dynamic clustering topic model (DCT).** This model captures the dynamic topic distribution of documents arriving during time period t in streams based on the content of the documents and the previous estimated distributions.

For the *Twitter-weak* dataset, the prior is correspond to Eqs. (5) to (6) in our proposed PADC model, for the *Twitter-strong* dataset, the prior is correspond to Eqs. (7) to (8) in PADC model, for the *News-weak* dataset, the prior is correspond to Eqs. (1) to (2) in PADC model, and for the *News-strong* dataset, the prior is correspond to Eqs. (3) to (4) in PADC model. The performance was computed by taking the average of these 10 experiments. Tables 5 and 6 depict document dynamic clustering performance in 3 unit time acquired by PADC, DCT, TTM, DMM, and DTM on the 4 data sets. The experimental results show that PADC model achieves better performance compared with the state-

Table 5. Comparison of the dynamic clustering performance on the Twitter-strong and Twitter-weak

Model	Twitter-weak	Twitter-strong
PADC	**0.623, 0.604, 0.601**	**0.639, 0.677, 0.684**
DCT	0.586, 0.577, 0.575	0.588, 0.620, 0.604
TTM	0.506, 0.501, 0.503	0.543, 0.549, 0.544
DMM	0.355, 0.357, 0.352	0.398, 0.394, 0.390
DTM	0.309, 0.304, 0.311	0.336, 0.334, 0.332

Table 6. Comparison of the dynamic clustering performance on the News-strong and News-weak

Model	News-weak	News-strong
PADC	**0.527, 0.513, 0.517**	**0.526, 0.521, 0.513**
DCT	0.462, 0.471, 0.473	0.469, 0.476, 0.465
TTM	0.376, 0.373, 0.372	0.395, 0.399, 0.391
DMM	0.331, 0.335, 0.342	0.353, 0.356, 0.357
DTM	0.293, 0.296, 0.291	0.317, 0.313, 0.312

Table 7. Number of clusters estimated on the four data sets

Dataset	PADC	DCT
Twitter-weak	4, 5, 8	3, 20, 13
Twitter-strong	3, 5, 8	3, 8, 9
News-weak	10, 13, 17	17, 24, 19
News-strong	5, 9, 14	13, 17, 25

of-the-art models. PADC model is robust and effective for dynamic document clustering problem in various of datasets.

Table 7 shows the estimated number of clusters of PADC and DCT on four real data sets in three unit time. From the experimental results, PADC can obtain a relative accurate estimation compared with DCT on four data sets.

5   Conclusion and Future Work

In this paper, we proposed a novel prior-adjusted dynamic document clustering approach namely, PADC, which is able to adjust the topic inheritance process according to two important of datasets characteristics, in particular, the interval between dataset time slides and the size of dataset time slides. We mainly study the 4 dataset scenarios in which the different interval between dataset time slides and the different size of dataset time slides, and our experiments on synthetic and real datasets show that PADC model acquires high dynamic clustering accuracy and reasonable scenarios of topic inheritance. The comparison between PADC model and state-of-the-art approaches indicates that PADC is robust and effective for document dynamic clustering.

As future work we intent to research how to model the dynamic clustering approach on multi-source data sets rather than single-source, and more inherited factors will be taken into account our approach to improve the performance of dynamic clustering with variable lineage of topics.

Acknowledgments. The work described in this paper is substantially supported by the National Natural Science Foundation of China (Grant No. U1836205), the National Natural Science Foundation of China (Grant No. 61462011), the Major Research Program of the National Natural Science Foundation of China (Grant No. 91746116), the Major Applied Basic Research Program of Guizhou Province (Grant No. JZ20142001), the Major Special Science and Technology Projects of Guizhou Province (Grant No. [2017]3002), and the Science and Technology Projects of Guizhou Province (Grant No. [2018]1035).

References

1. Begum, N., Ulanova, L., Wang, J., Keogh, E.: Accelerating dynamic time warping clustering with a novel admissible pruning strategy. In: Proceedings of the 21st ACM SIGKDD International Conference on Knowledge Discovery and Data Mining, pp. 49–58. ACM (2015)
2. Blei, D.M., Lafferty, J.D.: Dynamic topic models. In: Proceedings of the 23rd International Conference on Machine Learning, pp. 113–120. ACM (2006)
3. Blei, D.M., Ng, A.Y., Jordan, M.I.: Latent Dirichlet allocation. J. Mach. Learn. Res. **3**(Jan), 993–1022 (2003)
4. Chien, J.T., Lee, C.H., Tan, Z.H.: Latent Dirichlet mixture model. Neurocomputing **278**, 12–22 (2018). Recent Advances in Machine Learning for Non-Gaussian Data Processing. https://doi.org/10.1016/j.neucom.2017.08.029
5. Croft, W.B., Metzler, D., Strohman, T.: Search Engines: Information Retrieval in Practice, vol. 283. Addison-Wesley, Reading (2010)
6. Du, N., Farajtabar, M., Ahmed, A., Smola, A.J., Song, L.: Dirichlet-Hawkes processes with applications to clustering continuous-time document streams. In: Proceedings of the 21st ACM SIGKDD International Conference on Knowledge Discovery and Data Mining, pp. 219–228. ACM (2015)
7. Efron, M., Lin, J., He, J., De Vries, A.: Temporal feedback for tweet search with non-parametric density estimation. In: Proceedings of the 37th International ACM SIGIR Conference on Research and Development in Information Retrieval, pp. 33–42. ACM (2014)
8. He, Y., Lin, C., Gao, W., Wong, K.F.: Dynamic joint sentiment-topic model. ACM Trans. Intell. Syst. Technol. **5**(1), 6:1–6:21 (2014). https://doi.org/10.1145/2542182.2542188
9. Hofmann, T.: Probabilistic latent semantic indexing. SIGIR Forum **51**(2), 211–218 (2017). https://doi.org/10.1145/3130348.3130370
10. Huang, F., Zhang, S., Zhang, J., Yu, G.: Multimodal learning for topic sentiment analysis in microblogging. Neurocomputing **253**, 144–153 (2017). Learning Multimodal Data. https://doi.org/10.1016/j.neucom.2016.10.086

11. Injadat, M., Salo, F., Nassif, A.B.: Data mining techniques in social media: a survey. Neurocomputing **214**, 654–670 (2016). https://doi.org/10.1016/j.neucom.2016.06.045
12. Ishwaran, H., James, L.F.: Gibbs sampling methods for stick-breaking priors. J. Am. Stat. Assoc. **96**(453), 161–173 (2001)
13. Iwata, T., Watanabe, S., Yamada, T., Ueda, N.: Topic tracking model for analyzing consumer purchase behavior. IJCAI **9**, 1427–1432 (2009)
14. Iwata, T., Yamada, T., Sakurai, Y., Ueda, N.: Online multiscale dynamic topic models. In: Proceedings of the 16th ACM SIGKDD International Conference on Knowledge Discovery and Data Mining, pp. 663–672. ACM (2010)
15. Liang, S., de Rijke, M.: Burst-aware data fusion for microblog search. Inf. Process. Manag. **51**(2), 89–113 (2015)
16. Liang, S., Yilmaz, E., Kanoulas, E.: Dynamic clustering of streaming short documents. In: Proceedings of the 22nd ACM SIGKDD International Conference on Knowledge Discovery and Data Mining, pp. 995–1004. ACM (2016)
17. Liu, H., Ge, Y., Zheng, Q., Lin, R., Li, H.: Detecting global and local topics via mining Twitter data. Neurocomputing **273**, 120–132 (2018). https://doi.org/10.1016/j.neucom.2017.07.056
18. Porteous, I., Newman, D., Ihler, A., Asuncion, A., Smyth, P., Welling, M.: Fast collapsed gibbs sampling for latent Dirichlet allocation. In: Proceedings of the 14th ACM SIGKDD International Conference on Knowledge Discovery and Data Mining, pp. 569–577. ACM (2008)
19. Qi, S., Wang, F., Wang, X., Wei, J., Zhao, H.: Live multimedia brand-related data identification in microblog. Neurocomputing **158**, 225–233 (2015)
20. Teh, Y.W., Jordan, M.I., Beal, M.J., Blei, D.M.: Hierarchical Dirichlet processes. Publ. Am. Stat. Assoc. **101**(476), 1566–1581 (2006)
21. Vosecky, J., Jiang, D., Leung, K.W.T., Ng, W.: Dynamic multi-faceted topic discovery in Twitter. In: Proceedings of the 22nd ACM International Conference on Conference on Information & Knowledge Management, pp. 879–884. ACM (2013)
22. Wang, X., McCallum, A.: Topics over time: a non-Markov continuous-time model of topical trends. In: Proceedings of the 12th ACM SIGKDD International Conference on Knowledge Discovery and Data Mining, pp. 424–433. ACM (2006)
23. Wei, X., Croft, W.B.: LDA-based document models for ad-hoc retrieval. In: Proceedings of the 29th Annual International ACM SIGIR Conference on Research and Development in Information Retrieval, pp. 178–185. ACM (2006)
24. Wei, X., Sun, J., Wang, X.: Dynamic mixture models for multiple time-series. IJCAI **7**, 2909–2914 (2007)
25. Wu, L., Wang, D., Zhang, X., Liu, S., Zhang, L., Chen, C.W.: MLLDA: multi-level LDA for modelling users on content curation social networks. Neurocomputing **236**, 73–81 (2017). Good Practices in Multimedia Modeling. https://doi.org/10.1016/j.neucom.2016.08.114
26. Xianghua, F., Guo, L., Yanyan, G., Zhiqiang, W.: Multi-aspect sentiment analysis for chinese online social reviews based on topic modeling and hownet lexicon. Knowl.-Based Syst. **37**, 186–195 (2013)
27. Xiong, S., Wang, K., Ji, D., Wang, B.: A short text sentiment-topic model for product reviews. Neurocomputing **297**, 94–102 (2018). https://doi.org/10.1016/j.neucom.2018.02.034
28. Yan, X., Guo, J., Lan, Y., Cheng, X.: A biterm topic model for short texts. In: Proceedings of the 22nd International Conference on World Wide Web, pp. 1445–1456. ACM (2013)

29. Yin, J., Wang, J.: A Dirichlet multinomial mixture model based approach for short text clustering. In: Proceedings of the 20th ACM SIGKDD International Conference on Knowledge Discovery and Data Mining, pp. 233–242. ACM (2014)
30. Zhai, C., Lafferty, J.: A study of smoothing methods for language models applied to ad hoc information retrieval. In: ACM SIGIR Forum, vol. 51, pp. 268–276. ACM (2017)
31. Zhang, X., Chen, X., Chen, Y., Wang, S., Li, Z., Xia, J.: Event detection and popularity prediction in microblogging. Neurocomputing **149**, 1469–1480 (2015). https://doi.org/10.1016/j.neucom.2014.08.045
32. Zhao, W.X., et al.: Comparing Twitter and traditional media using topic models. In: Clough, P., et al. (eds.) ECIR 2011. LNCS, vol. 6611, pp. 338–349. Springer, Heidelberg (2011). https://doi.org/10.1007/978-3-642-20161-5_34
33. Zhong, S.: Semi-supervised Model-Based Document Clustering: A Comparative Study. Kluwer Academic Publishers, Hingham (2006)

An Improved Proof of the Closure Under Homomorphic Inverse of FCFL Valued in Lattice-Ordered Monoids

Haihui Wang, Luyao Zhao, and Ping Li[✉]

College of Mathematics and Information Science, Shaanxi Normal University,
Xi'an 710062, China
{wanghaihui,zhaoluyao,liping}@snnu.edu.cn

Abstract. We study fuzzy context-free grammars (FCFG), fuzzy context-free languages (FCFL) and fuzzy pushdown automata (FPDA) valued in lattice-ordered monoids. Inspired by the ideas of crisp cases, we get similar results, but some of them depend on commutative law. Particularly, we give two proofs of the closure under homomorphic inverse of FCFL when lattice-ordered monoids are commutative. Comparing with the classical method, we show that the improved proof is more efficient.

Keywords: Lattice-ordered monoid · Fuzzy context-free language · Fuzzy pushdown automata · Homomorphic inverse · Closure property

1 Introduction

The characterization of formal languages is a very important aspect of research in the classical computation theory. For example, regular languages can be characterized by some finite automata, regular expressions and regular grammars. However, the problem of vagueness and imprecision are frequently encountered in the study of natural languages. In order to represent the imprecision of natural languages, various fuzzy languages have been proposed. By introducing the concept of fuzziness into the structure of formal grammars, the concept of fuzzy automata was introduced by Santos [22,23] as early as in the late 1960s, and after the concept of fuzzy grammar was introduced by Lee and Zadeh [13,14]. There are amount of work on fuzzy languages, such as the relationships between fuzzy grammars and fuzzy languages, the relations of fuzzy automata and fuzzy language, and the algebraic properties of fuzzy languages [2,17,18]. As a further extension, Kim et al. put forward one type of L-fuzzy grammar based on the distributive lattice and Boolean lattice, another type of L-fuzzy grammar based on lattice-ordered monoid by assigning the element of lattice to the rewriting rules of a formal grammar [11]. Gerla also studied fuzzy grammars and recursively enumerable fuzzy languages, and proved that a fuzzy language can be

Supported by the National Natural Science Foundation of China under Grant 11301321, Grant 61673250, Grant 61672023.

© Springer Nature Singapore Pte Ltd. 2019
X. Sun et al. (Eds.): NCTCS 2019, CCIS 1069, pp. 64–75, 2019.
https://doi.org/10.1007/978-981-15-0105-0_5

generated by a fuzzy grammar if and only if it is recursively enumerable [6]. As one of the generators of fuzzy languages, fuzzy grammars have been used to solve some important questions such as intelligent interface design (Senay) [25], lexical analysis, clinical monitoring (Steimann and Adlassning) [26], neural networks (Giles et al.) [19], and pattern recognition (Depalma and Yau) [4]. It is known that the context-free grammars are powerful than regular grammars, also the classical pushdown automata and finite automata [9, 24]. Therefore, fuzzy pushdown automata and fuzzy context-free grammars were discussed in [3, 12, 16, 24, 27, 28] and automata theory based on residuated lattice-valued logic has been investigated recently [21, 28]. It is shown that their many properties are similar to classical pushdown automata, context-free grammars and context-free languages. In particular, we can get a much more efficient way to prove the closure under homomorphic inverse of fuzzy context-free languages.

In this paper, fuzzy context-free languages whose codomain forms a commutative lattice-ordered monoid $\mathcal{L}$ are studied. Furthermore, we show that the homomorphic inverse of fuzzy context-free languages are also fuzzy context-free languages by two ways, and the latter is much more efficient than the former by comparison.

The rest of the paper is arranged as follows. In Sect. 2, we introduce lattice-ordered monoids and give some examples. In Sect. 3, we discuss fuzzy context-free grammars, fuzzy context-free languages and their relationships. Section 4 studies the fuzzy pushdown automata valued in a lattice-ordered monoid $\mathcal{L}$. In Sect. 5, comparing with the classical cases, we give an improved proof of the closure under homomorphic inverse of fuzzy context-free languages, which increases the efficiency of operation. Finally, some conclusions are concerned in Sect. 6.

2 Lattice-Ordered Monoids

We first introduce the definition of lattice-ordered monoid and give some examples.

Definition 2.1. *Given a lattice $\mathcal{L}$, we use $\vee, \wedge$ to represent the supremum operation and infimum operation on $\mathcal{L}$, respectively. We need $\mathcal{L}$ to have the least and largest elements in the paper, which will be denoted as 0, 1, respectively. Assume that there is a binary operation $\bullet$ (we call it multiplication) on $\mathcal{L}$ such that $(\mathcal{L}, \bullet, e)$ is a monoid with identity $e \in \mathcal{L}$.*

We call $\mathcal{L}$ an po-monoid (some modification of the notion of partially ordered monoid in [5]) if it satisfies the following two conditions for any $a, b, x \in \mathcal{L}$,

(1) $\forall a \in \mathcal{L}, a \bullet 0 = 0 \bullet a = 0$,
(2) $a \leq b \Rightarrow a \bullet x \leq b \bullet x$ and $x \bullet a \leq x \bullet b$.

And we call $\mathcal{L}$ a lattice-ordered monoid or l-monoid if $\mathcal{L}$ is a po-monoid and it satisfies the distributive laws, i.e.,

(3) $\forall a, b, c \in \mathcal{L}, a \bullet (b \vee c) = (a \bullet b) \vee (a \bullet c)$ and $(b \vee c) \bullet a = (b \bullet a) \vee (c \bullet a)$.

Moreover, if $\mathcal{L}$ is a complete lattice, and it satisfies the following infinite distributive laws,

(4) $a \bullet (\bigvee_t b_t) = \bigvee_t (a \bullet b_t)$ and $(\bigvee_t b_t) \bullet a = \bigvee_t (b_t \bullet a)$,

where T is an index set, then we call $\mathcal{L}$ a quantale. If the distributive laws in (3) holds only for countable set $\{b_t\}_{t \in T}$ then $\mathcal{L}$ is called a countable lattice-ordered monoid.

We call the lattice-ordered monoid $(\mathcal{L}, \bullet, \vee)$ is commutative, if $a \bullet b = b \bullet a$ holds for any $a, b \in \mathcal{L}$. For a lattice-ordered monoid, we only concern the multiplication $\bullet$ and finite supremum operation $\vee$, in what follows, a lattice-ordered is denoted as $(\mathcal{L}, \bullet, \vee)$. If we deal with the subalgebra $\mathcal{L}_1$ of a lattice-ordered monoid$(\mathcal{L}, \bullet, \vee)$, it means that $\mathcal{L}_1$ is a nonempty subset of $\mathcal{L}$ and $\mathcal{L}_1$ is closed under the multiplication and finite supremum of $\mathcal{L}$. We do not concern with the infimum operation in $\mathcal{L}_1$.

The followings are discussed on the lattice-ordered monoid $\mathcal{L}$ without special instructions.

Example 2.1.

(1) Let $(\mathcal{L}, \wedge, \vee)$ be a distributive lattice, and let $\wedge = \bullet$, then $\mathcal{L}$ is a lattice-ordered monoid, and the identity of multiplication is 1.

(2) Let $(\mathcal{L}, \bullet, \vee)$ be a lattice-ordered monoid, the identity is e. We use $L(n)$ to denote all $n \times n$ matrices with values in $\mathcal{L}$. The multiplication, denote as $\circ$, is defined as $Sup - \bullet$ composition; and $\vee$ is the pointwise-$\vee$. That is, for two $n \times n$ matrices, $A = (a_{ij}), B = (b_{ij})$, with values in $\mathcal{L}$, let $A \circ B = C = (c_{ij})$, then $c_{ij} = \vee_{k=1}^n (a_{ik} \bullet b_{kj})$ and let $A \vee B = D = (d_{ij}), d_{ij} = a_{ij} \vee b_{ij}$. Then $(L(n), \circ, \vee)$ is also a lattice-ordered monoid, the identity is the diagonal-matrix $E = diag(e, \cdots, e)$, with e as the diagonal element. In general, the multiplication on $L(n)$ is not commutative, even if the multiplication on $\mathcal{L}$ is commutative.

(3) Let $\bullet$ be any uninorm on [0,1]. If $0 \bullet 1 = 0$, then$([0,1], \bullet, \vee)$ is a commutative lattice-ordered monoid. In particular, if $\bullet$ is a t-norm on [0,1], then $0 \bullet 1 = 0$, then $([0,1], \bullet, \vee)$ is a commutative lattice-ordered monoid with identity $e = 1$.

(4) Complete residuated lattices are special kinds of quantales, where its multiplication is commutative and the identity is the same as the largest element.

3   Fuzzy Grammars

In this section, we study fuzzy context-free grammars, fuzzy context-free languages and their relationships.

Definition 3.1. *Fuzzy grammar is a four tuple $G = (N, T, P, S)$, where N is a finite nonterminal alphabet and its elements are called variables; T is a finite terminal alphabet, $N \cap T = \emptyset$; $S \in N$ is a start symbol; P is a finite alphabet of fuzzy productions $u \to^\rho v$, where $u \in (N \bigcup T)^* N (N \bigcup T)^*$, $v \in (N \bigcup T)^*$, and $\rho \in \mathcal{L} - \{0\}$ represents the membership value of rewriting rule $u \to v$. We suppose S only appears in the left of fuzzy production $u \to v$.*

For $u_1, \cdots, u_n \in (N \bigcup T)^*$, if $u_1 \Rightarrow^{\rho_1} u_2 \Rightarrow^{\rho_2} \cdots \Rightarrow^{\rho_{n-1}} u_n$, then u_n called the derivation chain from u_1, denoted as $u_1 \Rightarrow^{\rho}_* u_n$, where $\rho = \rho_1 \bullet \rho_2 \bullet \cdots \bullet \rho_{n-1}$.

Definition 3.2. *The fuzzy language $L(G) : T^* \to \mathcal{L}$ generated by fuzzy grammar G is defined as, for any $\theta \in T^*$,*

$$L(G)(\theta) = \vee \{\rho | S \Rightarrow^{\rho}_* \theta\}.$$

Fuzzy grammar, also called type 0 grammar or fuzzy phrase grammar, it is further classified as follows.

Suppose that $G = (N, T, P, S)$ is a fuzzy grammar, then

(1) G is called fuzzy context-sensitive grammar (FCSG) or type 1 grammar, if for any $u \to^{\rho} v \in P$, there is $|u| \leq |v|$. And $L(G)$ is called fuzzy context-sensitive language (FCSL).
(2) G is called fuzzy context-free grammar (FCFG) or type 2 grammar, if for any $u \to^{\rho} v \in P$, there is $|u| \leq |v|$ and $u \in N$. And $L(G)$ is called fuzzy context-free language (FCFL).
(3) G is called fuzzy regular grammar (FRG) or type 3 grammar, if for any $u \to^{\rho} v \in P$, there is $u \in N$ and $v \in TB, B \in N \cup \{\varepsilon\}$, or $u = S, v = \varepsilon$. And $L(G)$ is called fuzzy context-free language (FRL).

In the paper, we only concern with the fuzzy context-free grammar.

Definition 3.3. *Suppose that $G = (N, T, P, S)$ is a fuzzy context-free grammar, then G is called fuzzy Chomsky normal form (FCNF) if for any production formula of G they have the form:*

$$A \to^{\rho} BC \ or \ A \to^{\rho} a \ or \ S \to^{\rho} \varepsilon,$$

where $A, B, C \in N, a \in T, \rho \in \mathcal{L} - \{0\}$.

Theorem 4.1. *For any fuzzy context-free grammar, there is an equivalent Chomsky normal form.*

4   Fuzzy Pushdown Automata

Now, we give the fuzzy pushdown automata valued in a lattice-ordered monoid $\mathcal{L}$.

Definition 4.1. *Fuzzy pushdown automata (FPDA) is a seven tuple $M = (Q, \Sigma, \Gamma, \delta, q_0, Z_0, F)$, where Q, Σ, Γ are nonempty finite sets, and they represent state sets, input alphabet, stack alphabet, respectively. $q_0 \in Q, Z_0 \in \Gamma$ represent initial state and start symbol. $F : Q \to \mathcal{L}$ is a fuzzy final state, $\delta : Q \times (\Sigma \cup \{\varepsilon\}) \times \Gamma \to F(Q \times \Gamma^*)$ which image is a finite fuzzy subset of $Q \times \Gamma^*$ is called fuzzy transition function.*

For any $p, q \in Q, \sigma \in \Sigma, Z \in \Gamma, \gamma \in \Gamma^*, \delta(q, \sigma, Z)(p, \gamma)$ means the possible degree that automata can enter next state p and the top stack letter Z transfer to γ when current state is q, the top stack letter is Z and the input symbol is σ. Similarly, $\delta(q, \varepsilon, Z)(p, \gamma)$ means the possible degree that the automata turn to the next state p and the top stack letter Z transfer to γ when current state is q, the top stack letter is Z and the input symbol is empty string. What's more, for any $(q, w, \gamma) \in Q \times \Sigma^* \times \Gamma^*$, it is called a instantaneous description of M and ID for short, it means that M is on the current state q, w is the untreated input string and M is focusing on the first character of w, the string in the stack is γ.

For $\alpha, \beta \in \Gamma^*$, β is the tail of α if there exists $\gamma \in \Gamma^*$ such that $\alpha = \gamma\beta$, denoted $\beta \leq \alpha$ and $\gamma = \alpha \setminus \beta$, $head(\beta)$ is the first character of β. Then we give the extension ∇ of δ as follows.

Definition 4.2. *Given a fuzzy pushdown automata* $M = (Q, \Sigma, \Gamma, \delta, q_0, Z_0, F)$, *we define the fuzzy relation* ∇ *in* $Q \times \Sigma^* \times \Gamma^*$ *as,*

$$\nabla((p, w, \beta), (q, v, \alpha)) = \begin{cases} \delta(p, \varepsilon, head(\beta))(q, \alpha \setminus tail(\beta)), & v = w, tail(\beta) \leq \alpha, \\ \delta(p, head(w), head(\beta)(q, \alpha \setminus tail(\beta)), & v = tail(w), tail(\beta) \leq \alpha, \\ 0, & otherwise. \end{cases}$$

∇^* is defined as the reflexive and transitive closure of ∇. The reflexive and transitive closure of fuzzy relation R in set Q is defined as $R^* = I \cup R \cup R \circ R \cup \cdots \cup R^n \cup \cdots$, where $\circ$ is $Sup - \bullet$ composition of fuzzy relations.

Definition 4.3. *Given a fuzzy pushdown automata* $M = (Q, \Sigma, \Gamma, \delta, q_0, Z_0, F)$, *for all* $\theta \in \Sigma^*$,
(1) *The fuzzy language accepted by M by final state is defined as,*

$$L(M)(\theta) = \bigvee_{q \in Q, \gamma \in \Gamma^*} [\nabla^*((q_0, \theta, Z_0), (q, \varepsilon, \gamma)) \bullet F(q)].$$

(2) *The fuzzy language accepted by M by empty stack is defined as,*

$$N(M)(\theta) = \bigvee_{q \in Q} \nabla^*((q_0, \theta, Z_0), (q, \varepsilon, \varepsilon)).$$

Theorem 4.1. *For an FPDA that accepts a fuzzy language L by empty stack, there exists another FPDA that accepts L by final states. And the vise versa.*

Corollary 4.1. *The fuzzy final states can be taken as crisp states when the FPDA accept fuzzy language by the final state.*

Theorem 4.2. *For a fuzzy language, it can be accepted by an FPDA iff the fuzzy language is FCFL.*

5   The Properties of Fuzzy Context-Free Language

this section, we investigate some properties of fuzzy context-free languages, and except the closure under homomorphic inverse, the proofs of other conclusions are similar to the classical conditions, so we don't give their proofs in this paper.

Definition 5.1. *Let f, g be any fuzzy context-free language on Σ^* that valued in a lattice-ordered monoid $\mathcal{L}$, and $a \in \mathcal{L}, \theta \in \Sigma^*$, then we defined the operations as follows:*

(1) The scalar product of a and f, denoted af and fa, is defined as, $(af)(\theta) = a \bullet f(\theta)$ and $(fa)(\theta) = f(\theta) \bullet a$.

(2) The union of f_1 and f_2, denoted $f_1 \cup f_2$, is defined as, $(f_1 \cup f_2)(\theta) = f_1(\theta) \vee f_2(\theta)$.

(3) The concatenation of f_1 and f_2, denoted $f_1 f_2$, is defined as, $(f_1 f_2)(\theta) = \bigvee_{\theta_1 \theta_2 = \theta} [f_1(\theta_1) \bullet f_2(\theta_2)]$. And then the concatenation operation satisfies the associative laws.

Definition 5.2. *Let Σ, Δ be finite nonempty character sets, $\phi : \Sigma \to F(\Delta^*)$ is a mapping, and ϕ is called fuzzy context-free substitution if $\phi(\sigma)$ is a fuzzy context-free language of Δ for any $\sigma \in \Sigma$.*

Given a mapping $\phi : \Sigma \to F(\Delta^*)$, then ϕ can be extended on Σ^* by the way as follows,

(1) $\phi(\varepsilon) = \{\frac{1}{\varepsilon}\}$;

(2) $\forall \theta \in \Sigma^*, \forall \sigma \in \Sigma, \phi(\theta\sigma) = \phi(\theta)\phi(\sigma)$.

Then we extend ϕ on $F(\Sigma^*)$, denoted as, $\phi : F(\Sigma^*) \to F(\Delta^*)$, and $\forall h \in F(\Sigma^*), \forall \omega \in \Delta^*$,

$$\phi(h)(\omega) = \bigvee_{\theta \in \Sigma^*} [h(\theta) \bullet \phi(\theta)(\omega)].$$

Definition 5.3. *A mapping $\phi : \Sigma \to F(\Delta^*)$ is called fuzzy homomorphism, if for any $\sigma \in \Sigma$, there exists unique $w \in \Sigma^*, a \in \mathcal{L} - \{0\}$ such that $\phi(\sigma) = \frac{a}{w}$. For the above mapping ϕ and $\sigma \in \Sigma$, we can define two mappings $\phi_1 : \Sigma \to \Delta^*$ and $\phi_2 : \Sigma \to \mathcal{L}$ as, $\phi_1(\sigma) = w$, $\phi_2(\sigma) = a$, respectively. Then ϕ_1 is the usual homomorphism, therefore, $\forall \sigma \in \Sigma, \phi(\sigma) = \frac{\phi_2(\sigma)}{\phi_1(\sigma)}$.*

Then we can extend ϕ_1 and ϕ_2 on Σ^* as, $\forall \theta = \sigma_1 \cdots \sigma_k \in \Sigma^*, \phi_1(\sigma_1 \cdots \sigma_k) = \phi_1(\sigma_1) \cdots \phi_1(\sigma_k)$, and $\phi_2(\sigma_1 \cdots \sigma_k) = \phi_2(\sigma_1) \bullet \cdots \bullet \phi_2(\sigma_k)$. Besides, we declare that $\phi_1(\varepsilon) = \varepsilon$, $\phi_2(\varepsilon) = e$.

Thus, we can extend ϕ on $F(\Sigma^*)$ as,

$$\forall g \in F(\Sigma^*), \quad \forall \omega \in \Delta^*, \quad \phi(g)(\omega) = \vee\{g(\theta) \bullet \phi_2(\theta) | \phi_1(\theta) = \omega\},$$

at the same time, we define the inverse mapping $\phi^{-1} : F(\Delta^*) \to F(\Sigma^*)$ as,

$$\forall h \in F(\Delta^*), \quad \forall \theta \in \Sigma^*, \quad \phi^{-1}(h)(\theta) = h(\phi_1(\theta)) \bullet \phi_2(\theta).$$

Theorem 5.1. *Any FCFL is still an FCFL under the fuzzy context-free substitution.*

By the extensive definition of fuzzy homomorphism, we know that fuzzy context-free homomorphism is a special type of fuzzy context-free substitution, so we can get the following conclusion.

Corollary 5.1. *Suppose that* $\phi : \Sigma \rightarrow F(\Delta^*)$ *is a fuzzy homomorphism, then* $\phi(f)$ *is an FCFL on* Δ *if* f *is an FCFL on* Σ.

That is to say, fuzzy context-free languages are closed under the fuzzy homomorphism, now we consider the fuzzy homomorphic inverse of fuzzy context-free languages in Definition 5.2. Suppose that f is a fuzzy context-free language on Δ, $\phi : \Sigma \rightarrow F(\Delta^*)$ is a fuzzy homomorphism, then f can be described by a fuzzy context-free grammar and accepted by a fuzzy pushdown automata. If $\phi^{-1}(f)$ is a fuzzy context-free language, there must be a fuzzy context-free grammar and a fuzzy pushdown automata corresponding to it. Here we only consider designing a fuzzy pushdown automata to accept it. Assume that M_2 is a fuzzy pushdown automata that accepts f. We need construct a fuzzy pushdown automata M_1 which can simulate the processing of M_2 to $\phi_1(a)$ when M_1 reads character a. But, we can't ignore the degree of the processing, that is $\phi_2(a)$. For one thing, we need to store $\phi_1(a)$ in the finite controller of M_1, and the degree of this process is $\phi_2(a)$. For another thing, $\phi_1(a)$ is a string, we need to simulate the process of M_2 to $\phi_1(a)$. Now we give two ways to complete the process, and the second solution is an improved way.

(1) We can use an empty move of M_1 to simulate the process of M_2 to each character of $\phi_1(a)$, and M_2 running $|\phi_1(a)|$ steps when it finished processing $\phi_1(a)$. This way is similar to the process of classical case, but what we should concern about is just the possible degree each transition of M_2.

(2) We can use an empty move of M_1 to simulate the process of M_2 to the whole string $\phi_1(a)$, then let M_1 remember the state and stack letter, which are M turns to after it finished processing $\phi_1(a)$. In the process, the possible transition degree of M_1 is just the possible degree that M_2 processed string $\phi_1(a)$.

Remark 5.1. *From the definition of* ∇^*, *it is clearly that if* $M = (Q, \Sigma, \Gamma, \delta, q_0, Z_0, F)$ *is a fuzzy pushdown automata,* $p, q \in Q, x, y \in \Sigma^*, \alpha, \beta \in \Gamma^*$, *then* $\nabla^*((q, x, \alpha), (p, y, \beta)) = \nabla^*((q, x\omega, \alpha), (p, y\omega, \beta))$ *holds for any* $\omega \in \Sigma^*$.

We will frequently use the fact to prove the following theorem.

Theorem 5.2. *If* $\mathcal{L}$ *is commutative, then the homomorphic inverse of FCFL is FCFL.*

Proof I. Let f be a fuzzy context-free language of Δ, $\phi : \Sigma \rightarrow F(\Delta^*)$ is a fuzzy homomorphism. Then there is a fuzzy pushdown automata $M_2 = (Q_2, \Delta, \Gamma, \delta_2, q_0, Z_0, \{q_f\})$ which accepts f, that is, for any θ in Δ^*, $f(\theta) = L(M_2)(\theta) = \bigvee_{\gamma \in \Gamma^*} \nabla_2^* ((q_0, \theta, Z_0), (q_f, \varepsilon, \gamma))$. Now we construct a fuzzy pushdown automata $M_1 = (Q_1, \Delta, \Gamma, \delta_1, [q_0, \varepsilon], Z_0, [q_f, \varepsilon])$ as,

(1) $Q_1 = \{[q, x] | \forall a \in \Sigma \bigcup \{\varepsilon\}, x$ is a suffix of $\phi_1(a)\}$, where the number of the suffix of $\phi_1(a)$ is finite, and Q_2 is a finite set, so Q_1 is finite.

(2) $\forall q \in Q_2, a \in \Sigma \bigcup \{\varepsilon\}, Z \in \Gamma, \delta_1([q, \varepsilon], a, Z)([q, \phi_1(a)], Z) = \phi_2(a)$.

(3) $\forall q, p \in Q_2, u \in \Sigma \cup \{\varepsilon\}, \gamma \in \Gamma^*, \delta_1([q, ux], \varepsilon, Z)([p, x], \gamma) = \delta_2(q, u, Z)(p, \gamma)$, and for other cases, $\delta_1 = 0$.

Therefore, if $\theta = \varepsilon$, then

$$
\begin{aligned}
L(M_1)(\varepsilon) &= \bigvee_{\gamma \in \Gamma^*} \nabla_1^*(([q_0, \varepsilon], \varepsilon, Z_0), ([q_f, \varepsilon], \varepsilon, \gamma)) \\
&= \bigvee_{\gamma \in \Gamma^*} \nabla_1(([q_0, \varepsilon], \varepsilon, Z_0), ([q_0, \phi_1(\varepsilon)], \varepsilon, Z_0)) \bullet \nabla_1(([q_0, \phi_1(\varepsilon)], \varepsilon, Z_0), ([q_f, \varepsilon], \varepsilon, \gamma)) \\
&= \bigvee_{\gamma \in \Gamma^*} \delta_1([q_0, \varepsilon], \varepsilon, Z_0)([q_0, \phi_1(\varepsilon)], Z_0) \bullet \delta_1([q_0, \phi_1(\varepsilon)], \varepsilon, Z_0)([q_f, \varepsilon], \gamma) \\
&= \bigvee_{\gamma \in \Gamma^*} \nabla_2^*((q_0, \phi_1(\varepsilon), Z_0), (q_f, \varepsilon, \gamma)) \bullet \phi_2(\varepsilon) \\
&= f(\phi_1(\varepsilon)) \bullet \phi_2(\varepsilon) \\
&= \phi^{-1}(f)(\varepsilon).
\end{aligned}
$$

Note that for any $\theta \in \Sigma^+$, let $\theta = a_1 \cdots a_n$, $a_i \in \Sigma$, $\phi_1(a_i) = b_{i1} \cdots b_{im}$, $i = 1, 2, ..., n, |\phi_1(a_i)| = m$, then $\phi_1(\theta) = \phi_1(a_1) \cdots \phi_1(a_n)$, $\phi_2(\theta) = \phi_2(a_1) \bullet \cdots \bullet \phi_2(a_n)$, therefore,

$$
\begin{aligned}
L(M_1)(\theta) &= \bigvee_{\gamma \in \Gamma^*} \nabla_1^*(([q_0, \varepsilon], \theta, Z_0), ([q_f, \varepsilon], \varepsilon, \gamma)) \\
&= \bigvee_{\gamma_i \in \Gamma^*, q_i \in Q_2} [\nabla_1(([q_0, \varepsilon], a_1 \cdots a_n, Z_0), ([q_0, \phi_1(a_1)], a_2 \cdots a_n, Z_0)) \\
&\quad \bullet \nabla_1(([q_0, \phi_1(a_1)], \varepsilon a_2 \cdots a_n, Z_0), ([q_1, \varepsilon], a_2 \cdots a_n, \gamma_1))] \\
&\quad \bullet [\nabla_1(([q_1, \varepsilon], a_2 \cdots a_n, \gamma_1), ([q_1, \phi_1(a_2)], a_3 \cdots a_n, \gamma_1)) \\
&\quad \bullet \nabla_1(([q_1, \phi_1(a_2)], \varepsilon a_3 \cdots a_n, \gamma_1), ([q_2, \varepsilon], a_3 \cdots a_n, \gamma_2))] \bullet \cdots
\end{aligned}
$$

$$
\begin{aligned}
&\quad \bullet [\nabla_1(([q_{n-1}, \varepsilon], a_n, \gamma_{n-1}), ([q_{n-1}, \phi_1(a_n)], \varepsilon, \gamma_{n-1})) \\
&\quad \bullet \nabla_1(([q_{n-1}, \phi_1(a_n)], \varepsilon, \gamma_{n-1}), ([q_f, \varepsilon], \varepsilon, \gamma))] \\
&= \bigvee_{\gamma, \gamma_i \in \Gamma^*, q_i \in Q_2} [\delta_1(([q_0, \varepsilon], a_1, Z_0)([q_0, \phi_1(a_1)], Z_0) \\
&\quad \bullet \delta_1([q_0, b_{11} \cdots b_{1m}], \varepsilon, Z_0)([q_1, \varepsilon], \gamma_1)] \bullet \cdots \\
&\quad \bullet [\delta_1([q_{n-1}, \varepsilon], a_n, \gamma_{n-1})([q_{n-1}, \phi_1(a_n)], \gamma_{n-1}) \\
&\quad \bullet \delta_1([q_{n-1}, b_{n1} \cdots b_{nm}], \varepsilon, \gamma_{n-1})([q_f, \varepsilon], \gamma)] \\
&= \bigvee_{\gamma, \gamma_i \in \Gamma^*, q_i, q_{ij} \in Q_2} [\phi_2(a_1) \bullet \delta_1([q_0, b_{11} \cdots b_{1m}], \varepsilon, Z_0)([q_{01}, b_{12} \cdots b_{1m}], \gamma_{01}) \\
&\quad \bullet \delta_1([q_{01}, b_{12} \cdots b_{1m}], \varepsilon, \gamma_{01})([q_{02}, b_{13} \cdots b_{1m}], \gamma_{02}) \bullet \cdots \\
&\quad \bullet \delta_1([q_{0(m-1)}, b_{1m}], \varepsilon, \gamma_{0(m-1)})([q_1, \varepsilon], \gamma_1)]
\end{aligned}
$$

$\bullet \cdots \bullet$

$\phi_2(a_n) \bullet [\delta_1([q_{n-1}, b_{n1} \cdots b_{nm}], \varepsilon, \gamma_{(n-1)})([q_{(n-1),1}, b_{n2} \cdots b_{nm}], \gamma_{(n-1),1})$

$\bullet \delta_1([q_{(n-1),1}, b_{n2} \cdots b_{nm}], \varepsilon, \gamma_{(n-1),1})([q_{(n-1),2}, b_{n3} \cdots b_{nm}], \gamma_{(n-1),2}) \bullet \cdots$

$\bullet \delta_1([q_{(n-1)(m-1)}, b_{nm}], \varepsilon, \gamma_{(n-1)(m-1)})([q_f, \varepsilon], \gamma)]$

$= \bigvee\limits_{\gamma, \gamma_i \in \Gamma^*, q_i, q_{ij} \in Q_2} [\delta_2(q_0, b_{11}, Z_0)(q_{01}, \gamma_{01}) \bullet \delta_2(q_{01}, b_{12}, \gamma_{01})(q_{02}, \gamma_{02}) \bullet \cdots$

$\bullet \delta_2(q_{0,(m-1)}, b_{1m}, \gamma_{0,(m-1)})(q_1, \gamma_1)]$

$\bullet \cdots \bullet$

$[\delta_2(q_{n-1}, b_{n1}, \gamma_{n-1})(q_{(n-1),1}, \gamma_{(n-1),1}) \bullet \delta_2(q_{(n-1),1}, b_{n2}, \gamma_{(n-1),1})(q_{(n-1),2}, \gamma_{(n-1),2}) \bullet \cdots$

$\bullet \delta_2(q_{(n-1)(m-1)}, b_{nm}, \gamma_{(n-1)(m-1)})(q_f, \gamma)]$

$\bullet [\phi_2(a_1) \bullet \cdots \bullet \phi_2(a_n)]$

$= \bigvee\limits_{\gamma, \gamma_i \in \Gamma^*, q_i \in Q_2} [\nabla_2^*((q_0, b_{11} \cdots b_{1m}, Z_0), (q_1, \varepsilon, \gamma_1))]$

$\bullet [\nabla_2^*((q_1, b_{21} \cdots b_{2m}, \gamma_{10}), (q_2, \varepsilon, \gamma_2))] \bullet \cdots$

$\bullet [\nabla_2^*((q_{n-1}, b_{n1} \cdots b_{nm}, \gamma_{n-1}), (q_f, \varepsilon, \gamma))]$

$\bullet [\phi_2(a_1) \bullet \cdots \bullet \phi_2(a_n)]$

$= \bigvee\limits_{\gamma, \gamma_i \in \Gamma^*, q_i \in Q_2} [\nabla_2^*((q_0, \phi_1(a_1), Z_0), (q_1, \varepsilon, \gamma_1))$

$\bullet \nabla_2^*((q_1, \phi_1(a_2), \gamma_1), (q_2, \varepsilon, \gamma_2)) \bullet \cdots$

$\bullet \nabla_2^*((q_{n-1}, \phi_1(a_n), \gamma_{n-1}), (q_f, \varepsilon, \gamma))]$

$\bullet [\phi_2(a_1) \bullet \cdots \bullet \phi_2(a_n)]$

$= \bigvee\limits_{\gamma, \gamma_i \in \Gamma^*, q_i \in Q_2} [\nabla_2^*((q_0, \phi_1(a_1)\phi_1(a_2) \cdots \phi_1(a_n), Z_0), (q_1, \varepsilon\phi_1(a_2) \cdots \phi_1(a_n), \gamma_1))$

$\bullet \nabla_2^*((q_1, \phi_1(a_2)\phi_1(a_3) \cdots \phi_1(a_n), \gamma_1), (q_2, \varepsilon\phi_1(a_3) \cdots \phi_1(a_n), \gamma_2)) \bullet \cdots$

$\bullet \nabla_2^*((q_{n-1}, \phi_1(a_n), \gamma_{n-1}), (q_f, \varepsilon, \gamma))] \bullet \phi_2(\theta)$

$= \bigvee\limits_{\gamma \in \Gamma^*} [\nabla_2^*((q_0, \phi_1(a_1) \cdots \phi_1(a_n), Z_0), (q_f, \varepsilon, \gamma))] \bullet \phi_2(\theta)$

$= f(\phi_1(\theta)) \bullet \phi_2(\theta)$

$= \phi^{-1}(f)(\theta).$

Thus, $L(M_1) = \phi^{-1}(f)$.

Proof II. Let f be a fuzzy context-free language of Δ, $\phi : \Sigma \rightarrow F(\Delta^*)$ is a fuzzy homomorphism. Then there is a fuzzy pushdown automata $M_2 = (Q_2, \Delta, \Gamma, \delta_2, q_0, Z_0, \{q_f\})$ which accepts f. Now we construct a fuzzy pushdown automata $M_1 = (Q_1, \Delta, \Gamma, \delta_1, [q_0, \varepsilon], Z_0, [q_f, \varepsilon])$ as,

(1) $Q_1 = \{[q, x] | \forall a \in \Sigma \bigcup \{\varepsilon\}, x = \phi_1(a)\}$, where the number of $\phi_1(a)$ is finite, and Q_2 is a finite set, so Q_1 is finite.
(2) $\forall q \in Q_2, a \in \Sigma \bigcup \{\varepsilon\}, Z \in \Gamma, \delta_1([q, \varepsilon], a, Z)([q, \phi_1(a)], Z) = \phi_2(a)$,
(3) $\forall q, p \in Q_2, \gamma \in \Gamma^*, \delta_1([q, \phi_1(a)], \varepsilon, Z)([p, \varepsilon], \gamma) = \nabla_2^*((q, \phi_1(a), Z), (p, \varepsilon, \gamma))$, and for other cases, $\delta_1 = 0$.

Now we prove that $L(M_1) = \phi^{-1}(f)$. First, if $\theta = \varepsilon$, then

$$
\begin{aligned}
L(M_1)(\varepsilon) &= \bigvee_{\gamma \in \Gamma^*} \nabla_1^*(([q_0, \varepsilon], \varepsilon, Z_0), ([q_f, \varepsilon], \varepsilon, \gamma)) \\
&= \bigvee_{\gamma \in \Gamma^*} \nabla_1(([q_0, \varepsilon], \varepsilon, Z_0), ([q_0, \phi_1(\varepsilon)], \varepsilon, Z_0)) \bullet \nabla_1(([q_0, \phi_1(\varepsilon)], \varepsilon, Z_0), ([q_f, \varepsilon], \varepsilon, \gamma)) \\
&= \bigvee_{\gamma \in \Gamma^*} \nabla_2^*((q_0, \phi_1(\varepsilon), Z_0), (q_f, \varepsilon, \gamma)) \bullet \phi_2(\varepsilon) \\
&= f(\phi_1(\varepsilon)) \bullet \phi_2(\varepsilon) \\
&= \phi^{-1}(f)(\varepsilon).
\end{aligned}
$$

Note that for any $\theta \in \Sigma^+$, let $\theta = a_1 \cdots a_n$, $a_i \in \Sigma$, $i = 1, 2, ..., n$, $\phi_1(\theta) = \phi_1(a_1) \cdots \phi_1(a_n)$, $\phi_2(\theta) = \phi_2(a_1) \bullet \cdots \bullet \phi_2(a_n)$, then

$$
L(M_1)(\theta) = \bigvee_{\gamma \in \Gamma^*} \nabla_1^*(([q_0, \varepsilon], \theta, Z_0), ([q_f, \varepsilon], \varepsilon, \gamma))
$$

$$
= \bigvee_{\gamma_i \in \Gamma^*, q_i \in Q_2} [\nabla_1(([q_0, \varepsilon], a_1 \cdots a_n, Z_0), ([q_0, \phi_1(a_1)], a_2 \cdots a_n, Z_0))
$$

$$
\bullet \nabla_1(([q_0, \phi_1(a_1)], \varepsilon a_2 \cdots a_n, Z_0), ([q_1, \varepsilon], a_2 \cdots a_n, \gamma_1))]
$$

$$
\bullet [\nabla_1(([q_1, \varepsilon], a_2 \cdots a_n, \gamma_1), ([q_1, \phi_1(a_2)], a_3 \cdots a_n, \gamma_1))
$$

$$
\bullet \nabla_1(([q_1, \phi_1(a_2)], \varepsilon a_3 \cdots a_n, \gamma_1), ([q_2, \varepsilon], a_3 \cdots a_n, \gamma_2))] \bullet \cdots
$$

$$
\bullet [\nabla_1(([q_{n-1}, \varepsilon], a_n, \gamma_{n-1}), ([q_{n-1}, \phi_1(a_n)], \varepsilon, \gamma_{n-1}))
$$

$$
\bullet \nabla_1(([q_{n-1}, \phi_1(a_n)], \varepsilon, \gamma_{n-1}), ([q_f, \varepsilon], \varepsilon, \gamma))]
$$

$$
= \bigvee_{\gamma, \gamma_i \in \Gamma^*, q_i \in Q_2} [\delta_1([q_0, \varepsilon], a_1, Z_0)([q_0, \phi_1(a_1)], Z_0)
$$

$$
\bullet \delta_1([q_0, \phi_1(a_1)], \varepsilon, Z_0)([q_1, \varepsilon], \gamma_1)]
$$

$$
\bullet [\delta_1([q_1, \varepsilon], a_2, \gamma_1)([q_1, \phi_1(a_2)], \gamma_1)
$$

$$
\bullet \delta_1([q_1, \phi_1(a_2)], \varepsilon, \gamma_1)([q_2, \varepsilon], \gamma_2)] \bullet \cdots
$$

$$
\bullet [\delta_1([q_{n-1}, \varepsilon], a_n, \gamma_{n-1})([q_{n-1}, \phi_1(a_n)], \gamma_{n-1})
$$

$$
\bullet \delta_1([q_{n-1}, \phi_1(a_n)], \varepsilon, \gamma_{n-1})([q_f, \varepsilon], \gamma)]
$$

$$
= \bigvee_{\gamma, \gamma_i \in \Gamma^*, q_i \in Q_2} [\nabla_2^*((q_0, \phi_1(a_1), Z_0), (q_1, \varepsilon, \gamma_1))
$$

$$
\bullet \nabla_2^*((q_1, \phi_1(a_2), \gamma_1), (q_2, \varepsilon, \gamma_2)) \bullet \cdots
$$

$$
\bullet \nabla_2^*((q_{n-1}, \phi_1(a_n), \gamma_{n-1}), (q_f, \varepsilon, \gamma))]
$$

$$
\bullet [\phi_2(a_1) \bullet \cdots \bullet \phi_2(a_n)]
$$

$$
= \bigvee_{\gamma, \gamma_i \in \Gamma^*, q_i \in Q_2} [\nabla_2^*((q_0, \phi_1(a_1)\phi_1(a_2) \cdots \phi_1(a_n), Z_0), (q_1, \varepsilon\phi_1(a_2) \cdots \phi_1(a_n), \gamma_1))
$$

$$
\bullet \nabla_2^*((q_1, \phi_1(a_2)\phi_1(a_3) \cdots \phi_1(a_n), \gamma_1), (q_2, \varepsilon\phi_1(a_3) \cdots \phi_1(a_n), \gamma_2)) \bullet \cdots
$$

$$
\bullet \nabla_2^*((q_{n-1}, \phi_1(a_n), \gamma_{n-1}), (q_f, \varepsilon, \gamma))] \bullet \phi_2(\theta)
$$

$$= \bigvee_{\gamma \in \Gamma^*} [\nabla_2^*((q_0, \phi_1(a_1) \cdots \phi_1(a_n), Z_0), (q_f, \varepsilon, \gamma))] \bullet \phi_2(\theta)$$
$$= f(\phi_1(\theta)) \bullet \phi_2(\theta)$$
$$= \phi^{-1}(f)(\theta).$$

Thus, $L(M_1) = \phi^{-1}(f)$.

Clearly, the second solution is much more efficient than the first one. Let $|Q_2|, |\Sigma|$ represent the number of their elements, respectively. For any $a_i \in \Sigma, \phi_1(a_i) \in \Delta^*$, $|\phi_1(a_i)|$ represents the number of its suffixes. Then we need $m = |Q_2| \times |\Sigma| \times \prod_{a_i \in \Sigma} |\phi_1(a_i)|$ steps in the proof I, while we just need $n = |Q_2| \times |\Sigma|$ steps in the proof II, that is, $n \ll m$, which means that the improved proof has greatly improved the efficiency of operation. From two different proofs above, it can be seen that the construction of new state sets is a particularly essential step and the key to improve the efficiency.

6   Conclusion

In this paper, we have studied fuzzy context-free grammars (FCFG), fuzzy context-free languages(FCFL) and fuzzy pushdown automata(FPDA) whose condomain forms a lattice-ordered monoid $\mathcal{L}$. Some important conclusions are obtained. Particularly, if $\mathcal{L}$ is commutative, we show an improved proof of the closure under homomorphic inverse of fuzzy context-free languages, which has greatly improved the efficiency of the operation by comparing with the classical method. Undoubtedly, much more work remains to be completed along this line. There may be more special properties of fuzzy context-free languages on lattice-ordered monoids.

References

1. Alasdair, U.: Balbes Raymond and Dwinger Philip. Distributive lattices. University of Missouri Press, Columbia 1974, xiii + 294 pp. J. Symb. Logic **42**, 294–588 (1997)
2. Asveld, P.R.J.: Algebraic aspects of families of fuzzy languages. Theoret. Comput. Sci. **293**(2), 417–445 (2003)
3. Asveld, P.R.J.: Fuzzy context-free languages part 1: generalized fuzzy context-free grammars. Theoret. Comput. Sci. **347**(1), 167–190 (2005)
4. Depalma, G.F., Yau, S.S.: Fractionally fuzzy grammars with application to pattern recognition. In: Fuzzy Sets & Their Applications to Cognitive & Decision Processes, pp. 329–351 (1975)
5. Dilworth, R.P., Birkhoff, G.: Lattice theory. Bull. Am. Math. Soc. **56**, 204–206 (1950)
6. Gerla, G.: Fuzzy grammars and recursively enumerable fuzzy languages. Inf. Sci. **60**(1C2), 137–143 (1992)
7. Guo, X.: Grammar theory based on lattice-ordered monoid. Fuzzy Sets Syst. **160**(8), 1152–1161 (2009)
8. Guo, X.: A comment on automata theory based on complete residuated lattice-valued logic: pushdown automata (2012)

9. Jiang, Z., Jiang, S.: Formal languages and automata. Tsinghua University Press (2003)
10. Jin, J., Li, Q.: Fuzzy grammar theory based on lattices. Soft Comput. **16**(8), 1415–1426 (2012)
11. Kim, H.H., Mizumoto, M., Toyoda, J., Tanaka, K.: L -fuzzy grammars. Inf. Sci. **8**(2), 123–140 (1975)
12. Lan, S.: A machine accepted fuzzy context-free languages and fuzzy pushdown automata. BUSEFAL **49**(6), 67–72 (1992)
13. Lee, E.T., Zadeh, L.A.: Note on fuzzy languages. Inf. Sci. **1**(4), 421–434 (1969)
14. Lee, E.T., Zadeh, L.A.: Fuzzy languages and their acceptance by automata. In: Fourth Princeton Conference Information Science and System, no. 2, pp. 399–410 (1970)
15. Li, P., Li, Y.M.: Algebraic properties of la-languages. Inf. Sci. **176**(21), 3232–3255 (2006)
16. Li, Y., Li, P.: Fuzzy computing theory. Science Press (2016)
17. Li, Y., Pedrycz, W.: Fuzzy finite automata and fuzzy regular expressions with membership values in lattice-ordered monoids. Fuzzy Sets Syst. **156**(25), 68–92 (2005)
18. Li, Z., Li, P., Li, Y.: The relationships among several types of fuzzy automata. Inf. Sci. **176**(15), 2208–2226 (2006)
19. Omlin, C.W., Giles, C.L.: Equivalence in knowledge representation: automata, recurrent neural networks, and dynamical fuzzy systems. Proc. IEEE **87**(9), 1623–1640 (1999)
20. Pan, H., Song, F., Cao, Y., Qian, J.: Fuzzy pushdown termination games. IEEE Trans. Fuzzy Syst. **27**(15), 760–774 (2019)
21. Qiu, D.: Automata theory based on complete residuated latticed-valued logic(i). Sci. China (Series E) **33**(10), 137–146 (2003)
22. Santos, E.: Fuzzy automata and languages. Inf. Sci. **10**(3), 193–197 (1976)
23. Santos, E.S.: Maximin automata. Inf. Control **13**(4), 363–377 (1968)
24. Schtzenberger, M.: On context-free languages and push-down automata. Inf. Control **6**(3), 246–264 (1963)
25. Senay, H.: Fuzzy command grammars for intelligent interface design. IEEE Trans. Syst. Man Cybern. **22**(5), 1124–1131 (1992)
26. Steimann, F., Adlassnig, K.P.: Clinical monitoring with fuzzy automata. Fuzzy Sets Syst. **61**(1), 37–42 (1994)
27. Wang, H., Qiu, D.: Computing with words via turing machines: a formal approach. IEEE Trans. Fuzzy Syst. **11**(6), 742–753 (2003)
28. Xing, H.: Fuzzy pushdown automata. Fuzzy Sets Syst. **158**(13), 1437–1449 (2007)
29. Xing, H., Qiu, D.: Automata theory based on complete residuated lattice-valued logic: a categorical approach. Fuzzy Sets Syst. **160**(16), 2416–2428 (2009)

Socially-Attentive Representation Learning for Cold-Start Fraud Review Detection

Qiaobo Da[1,2], Jieren Cheng[1,2(✉)], Qian Li[3], and Wentao Zhao[4]

[1] Key Laboratory of Internet Information Retrieval of Hainan Province,
Hainan University, Haikou, China
cjr@hainu.edu.cn
[2] School of Computer and Cyberspace Security, Hainan University, Haikou, China
[3] Global Big Data Technologies Centre, University of Technology Sydney,
Sydney, Australia
[4] College of Computer, National University of Defense Technology, Changsha, China

Abstract. Fraud reviews consist one of the most serious issues in cyberspace, which dramatically damage users' decisions yet have great challenges to be detected. Accordingly, effectively detecting fraud reviews is becoming a critical task for cybersecurity. Although various efforts have been put on fraud review detection, they may fail in the case of cold-start where a review is posted by a new user who just pops up on social media. The main reason lies in lacking sufficient historical information of the new user. Recently, limited research has been conducted on fraud review detection with the cold-start problem, in which, however, advanced methods either ignore complex collaborative review manipulations or overlook fraud-related characteristics. As a result, they may easily be deceived by camouflage fraudsters and have low detection precision. This paper presents a novel socially-attentive user representation learning method for fraud review detection with the cold-start problem, namely SATURN, which leverages the fraud-related user reviewing behavior with comprehensive user social couplings for cold-start fraud review detection. SATURN jointly embeds user-item-attitude-review entities relations, explicit and implicit hierarchical social couplings, and fraud-related information into a user vector representation space, in which the fraud-related representation of a new user can be effectively inferred according to the learned socially-attentive entities relation. Subsequently, SATURN effectively detects fraud reviews with the cold-start problem in its learned representation space. Extensive experiments on four large real-world data sets demonstrate SATURN significantly outperforms three state-of-the-art and two baseline competitors in terms of both general and cold-start fraud review detection tasks.

Keywords: Fraud review detection · Cold-start ·
Socially-attentive representation · Coupling relations learning

© Springer Nature Singapore Pte Ltd. 2019
X. Sun et al. (Eds.): NCTCS 2019, CCIS 1069, pp. 76–91, 2019.
https://doi.org/10.1007/978-981-15-0105-0_6

1 Introduction

Nowadays, cyberspace brings great convenience for people with its rich information resources, in which the reviews posted by users heavily affect people's decision. However, the reliability of these reviews have becoming a serious issue with the emerging of the fraudsters who write fraud reviews to confuse honest users for great business values and reputations. For example, as reported by a recent research[1], Amazon's fake review problem is getting worse than ever in 2017. These fraud reviews dramatically damage users' decisions and are ubiquitous in cyberspace with various patterns. Accordingly, effectively detecting fraud reviews is becoming a critical task for cybersecurity.

Currently, various efforts have been put on fraud review detection and achieved significant progress. Typical fraud review detection methods mainly focus on review content. They design and extract a variety of linguistic features (n-gram, POS, etc.) from review content to identify fraud reviews [21]. Although review content provides certain evidence of fraud information, it can be manipulated by fraudsters. As a result, the content-based fraud review detection methods may easily be deceived by camouflage fraudsters who manipulate the content of fraud reviews similar to that of honest reviews. Therefore, it is unreliable to detect fraud reviews only based on review content [6]. To tackle the above problem, more advanced methods model user behavior for fraud review detection and show promising performance [27]. However, these methods require a large amount of historical data per each user to extract behavioral features. They may fail to extract the behavior features when facing the *cold-start* problem, i.e. *a new user just posted a new review*, because of lacking sufficient historical information of the new user.

Limited recent research has been conducted on fraud review detection with the cold-start problem. The first method to detect fraud reviews with the cold-start problem is proposed by [26] and followed by [28]. The methods in [26] and [28] both learn user behavior from the relations between users, items and reviews, namely entities relation, and embed such user behavior into a review vector representation space for fraud review detection. Compared with other behavior-based fraud review detection methods, these two methods are able to handle the cold-start cases because they model user behavior from review content, which is always available, instead of users' historical information. However, they ignore complex collaborative review manipulations that camouflage the content of fraud reviews like that of honest reviews. As a result, they may not be reliable to detect camouflaged fraud reviews as the other content-based methods discussed above. More recently, the method proposed by [14] solves this problem via jointly embedding the entities relations and user social relations into a user vector representation space. Different from the methods in [26] and [28], this method identifies whether a review posted by a user is fraud according to the reliability of other users who are similar to the user in the learned user

---

[1] https://www.forbes.com/sites/emmawoollacott/2017/09/09/exclusive-amazons-fake-review-problem-is-now-worse-than-ever.

representation space, which gains large performance improvement. It should be noted that this method only captures a piece of user social relations, i.e. user co-reviewing, but overlooks other complex social couplings, such as the same attitude and similar preferences, which may also reflect potential user collaborations. Besides, this method does not consider fraud-related information in the entities relations learning process, which may decrease the fraud review detection precision especially in the cold-start cases.

To address the above problems, in this paper, we propose a novel socially-attentive user representation learning method for fraud review detection with the cold-start problem, namely SATURN. The proposed SATURN embeds the user-item-attitude-review entities relations to tackle the cold-start problem and leverages the social relations to alleviate the effects of camouflaged review. To comprehensively capture the social relations, SATURN hierarchically learns and embeds both explicit and implicit user social couplings. Subsequently, the potential user collaborative manipulations can be revealed from different aspects. To preserve the fraud-related information, SATURN jointly embeds the entities relations, the social couplings, and the fraud-related information into a user vector representation space, in which the fraud-related representation of a new user can be effectively inferred via the learned socially-attentive entities relations.

Specifically, SATURN models the entities relations as that a *review* is written by a *user* for an *item* with an *attitude*. SATURN further hierarchically learns the *explicit user social couplings* that reflected by users' co-occurred social activities and the *implicit social couplings* built on users' similar demographics to form the comprehensive social relations. Then, SATURN optimizes the representations of users, items, attitudes and reviews to follow the entities relations and alignment with the fraud-related information. In this process, SATURN seamlessly integrates the social relations with the entities relations via adjusting the representation of a user by the representations of the user's neighbors per their social relations. After that, SATURN detects fraud reviews based on its learned representations of users, items, attitudes, and reviews. For a new user, SATURN infers the user representation according to the learned socially-attentive entities relations. In this way, SATURN enables an effective cold-start fraud review detection.

Accordingly, this paper makes three major contributions:

- **We propose a novel socially-attentive user representation learning method, SATURN, catering for cold-start fraud review detection.** SATURN simultaneously embeds the entities relations, the social relations, and the fraud-related information into a socially-attentive user representation, and thus, the comprehensively representation is more reliable for cold-start fraud review detection.
- **We propose a novel coupling learning method to comprehensively capture user social relations.** The coupling learning method hierarchically learns both the explicit user social couplings and implicit user social couplings to provide comprehensive evidence for fraud review detection.

– **We propose a novel method to seamlessly integrate user social couplings into user representations.** The proposed integration method implements an attention mechanism to optimize coupled users with similar representations, which enables an efficient yet effective socially-attentive representation learning.

Extensive experiments on four large real-world data sets demonstrate SATURN significantly outperforms four state-of-the-art and two baseline competitors in terms of both general and cold-start fraud review detection tasks.

2 Related Work

2.1 Fraud Review Detection

Currently, cybersecurity is an important problem and has attracted increasing attention [1–4,15]. Among these, fraud review detection is an emerging and challenging task. Fraud review detection was firstly researched by [9]. At that stage, most research were concentrate on extracting linguistic features from review content. However, they could not deal with real-life fraud reviews via mining and analyzing linguistic features because reviews could be easily manipulated by fraudsters [8,12,22]. In another words, the content-based fraud review detection methods could be deceived by camouflage fraudsters. As a result, more advanced methods model user behavior for fraud review detection [11–13]. Also, concentrating on behavior features is more effective than concentrating on linguistic feature for fraud review detection, which was been proved by [20]. Thus, the key point of most research transfers to extracting behavior features.

2.2 Cold-Start Problem

Recently, limited efforts have focused on fraud review detection with the cold-start problem. The first work on this topic is [26]. This method models user behavior as the relations between users, items and reviews to address the lacking of historical information in cold-start cases. Following this work, the method proposed by [28] further involves attribute and domain knowledge to enhance the user-item-review relation, which leads to a better cold-start fraud review detection performance. Although the above methods consider the user-item-review relation, they finally embeds the relation into review representations for fraud review detection. In other words, only review content has been used as the evidence for fraud identification. However, review content is easy to be manipulated, thus, the fraud reviews may be camouflaged as honest reviews by fraudsters [7]. As a result, these methods may fail in detecting fraud reviews where fraudsters are widely exists in real world. Besides, they ignore the user social relations which may reflect the collaborative manipulation. Later, the work in [14] detects cold-start fraud review from a novel aspect. Specifically, it jointly embeds the user-item-review entities relations and user social relations into a user representation space, and identifies a fraud review according to its posted user in that representation space. Although achieves significantly better performance, this method only captures the user co-reviewing relation but overlooks other complex social couplings, such as the same attitude and similar preferences. In addition, this method does not consider fraud-related information in the entities relations learning process, which may decrease the fraud review detection precision especially in the cold-start cases.

In this paper, we simultaneously captures the entities relation, comprehensive social couplings and fraud-related information to form the socially-attentive user representation for cold-start fraud review detection. As a result, we not only alleviate the effect of camouflaged reviews but also preserve the fraud-related information for more accurate fraud review detection.

3   Proposed Method

3.1   User Representation Learning Architecture

The architecture of the proposed SATURN model is shown in Fig. 1. It consists (1) *entities relations embedding* and (2) *social couplings learning*. Specially, in the entities relations embedding part, SATURN embeds the relations between users, items, attitude and reviews into their vector representations. In the social couplings learning part, it comprehensively model both explicit and implicit user social couplings from users' co-reviewing attitude and users' demographic information. SATURN integrates these two parts by a socially attentive learning method in which the fraud-related information is adopted to guide the learning process. Finally, it uses the learned representations to detect fraud reviews.

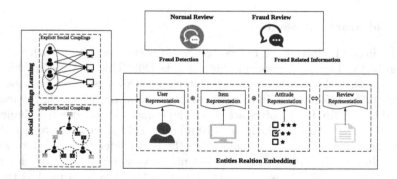

Fig. 1. SATURN architecture

3.2   Entities Relation Embedding

Inspired by the entities relation embedding method in [14,26,28], SATURN embeds the entities relation between users, items, and reviews. Different from the above methods [14, 26, 28], SATURN further consider the impacts of a user's attitude to an item when writing a review, which is reflected by the corresponding rating score. Specifically, SATURN assumes a review is wrote by a user to an item with a certain attitude, and this relation should be preserved and reflected in the SATURN's representation space.

Given an online review data set S, SATURN learns vectors $\mathbf{v}_u$, $\mathbf{v}_o$, $\mathbf{v}_r$, and $\mathbf{v}_s$ as the representation of the entities in a reviewing activity $\nu := <u, o, r, s> \in S$, where u refers to a user, o refers to an item, review r refers to a review, and s refers to a rating score. It embeds the entities relation between u, o, r and s in the representation space by optimizing the following objective function:

$$
\begin{aligned}
\min_{V, \Phi} & \sum_{i=1}^{n_\nu} \sum_{y=\{0,1\}} \mathbf{1}[y_i = y] \log q_i \\
& + \sum_{\nu \in S} \sum_{\nu' \notin S} \gamma \max\{0, 1 + \|\mathbf{v}_u + \mathbf{v}_o + \mathbf{v}_s - \mathbf{v}_r\|^2 \\
& \qquad\qquad\qquad - \|\mathbf{v}_{u'} + \mathbf{v}_{o'} + \mathbf{v}_{s'} - \mathbf{v}_{r'}\|^2\}, \\
s.t. \ \ \gamma &= \begin{cases} 1 & u = u' \\ 0 & u \neq u', \end{cases} \\
q_i &= \mathrm{softmax}(\mathbf{w} D_{\mathbf{p}}([\mathbf{v}_u, \mathbf{v}_o, \mathbf{v}_s, \mathbf{v}_r]) + b), \quad <u, o, s, r> \in \nu_i, \\
\mathbf{v}_r &= t_\omega(r),
\end{aligned}
\tag{1}
$$

where y_i is the ground-truth label of ν_i, $D_{\mathbf{p}}(\cdot)$ is a fully-connected network with parameter $\mathbf{p}$, $V = \{\mathbf{V}_u, \mathbf{V}_o, \mathbf{V}_s\}$ is the set of users', items', and rating scores' representations, $\Phi = \{\omega, \mathbf{p}, \mathbf{w}, b\}$ is the set of model parameters, $t_\omega(\cdot)$ refers to a text embedding neural network with parameters ω, $\max\{\cdot\}$ returns the maximum value in a set, and $\mathbf{V}_u$, $\mathbf{V}_o$, $\mathbf{V}_s$ are the representation matrices of user, item, rating score, respectively. In this paper, SATURN implements the text embedding neural network $t_\omega(\cdot)$ as the *universal transformer* network [5].

3.3 Social Couplings Learning

SATURN hierarchically learns both explicit and implicit social couplings between users via a two-steps approach. Here, the explicit social couplings refer to the coupling relationships reflected by users' co-occurred social activities, i.e. their reviewing activities to the same items. The implicit social couplings refer to the potential coupling relationships built on users' similar demographics.

In the first step, SATURN builds a user feature matrix $\mathbf{M}$ with size $n_u \times n_f$ where each row corresponds to a user, and each column corresponds to a feature. To capture the explicit couplings, SATURN uses the rating scores to build the user feature matrix. Specifically, SATURN assigns the value M_{ij}^{exp} in the user explicit feature matrix $\mathbf{M}^{exp}$, which corresponds to the i-th user u_i and the j-th item o_j, as the rating score s_{ij} given by u_i to o_j. If u_i does not review item o_j, SATURN sets the corresponding value M_{ij} as 0. The rationale is that the constructed explicit feature matrix reflects not only the items a user reviewed but also the attitude similarity of different users for the same item, which cannot be captured by the density subgraph mining method used in [14]. To capture the implicit couplings, SATURN adopts the demographic information to build to user feature matrix as $\mathbf{M}^{imp}$ where each column is a demographic feature, such as occupation, education, age and salary. These demographic features indicate the potential couplings between users. For example, students may prefer general textbooks, but data scientists may be more interested in books about analytics.

In the second step, SATURN follows the hierarchical coupling learning method in [29] to learn the complex coupling relationships between users, because most of the values in $\mathbf{M}^{exp}$ and $\mathbf{M}^{imp}$ belong to categorical value. For a value M_{ij}, SATURN first learns its intra-attribute couplings representation as follows,

$$c^{Ia}(M_{ij}) = \left[\frac{|g(M_{ij})|}{n_u}\right]^{\top},\tag{2}$$

where $g(M_{ij})$ returns a set of users that hold the feature value M_{ij}, and $|\cdot|$ counts the size of a set. The intra-attribute couplings capture the relations of different users per the same item. Further, SATURN learns the inter-attribute couplings representation for the value M_{ij} as follows,

$$c^{Ie}(M_{ij}) = \left[p(M_{ij}|M_1^*), \cdots, p(M_{ij}|M_k^*), \cdots, p(M_{ij}|M_{n_*}^*)\right]^{\top},\tag{3}$$

where M^* refers to the set of categorical values in all attributes except the j-th attribute, and $p(M_{ij}|M_k^*)$ is the co-occurrence frequency of value M_{ij} and M_k^* in the user feature that can be calculated as follows,

$$p(M_{ij}|M_k^*) = \frac{|g(M_{ij}) \cap g(M_k^*)|}{|g(M_k^*)|}.\tag{4}$$

The inter-attribute couplings capture the in-depth relations of different users per different items. Then, SATURN concatenates the intra- and inter-coupling representation to form the unified coupling representation of the user u_i:

$$c(u_i) = \left[c^{Ia}(M_{i1})^{\top}, \cdots, c^{Ia}(M_{in_f})^{\top}, c^{Ie}(M_{i1})^{\top}, \cdots, c^{Ie}(M_{in_f})^{\top}\right]^{\top}.\tag{5}$$

Finally, SATURN adopts linear kernel on the couplings representation space to construct the social coupling matrix of users, which can be formalized as follows,

$$\mathbf{C} = \begin{bmatrix} c(u_1)^{\top}c(u_1) & c(u_1)^{\top}c(u_2) & \cdots & c(u_1)^{\top}c(u_{n_u}) \\ c(u_2)^{\top}c(u_1) & c(u_2)^{\top}c(u_2) & \cdots & c(u_2)^{\top}c(u_{n_u}) \\ \vdots & \vdots & \ddots & \vdots \\ c(u_{n_u})^{\top}c(u_1) & c(u_{n_u})^{\top}c(u_2) & \cdots & c(u_{n_u})^{\top}c(u_{n_u}) \end{bmatrix}.\tag{6}$$

3.4 Socially-Attentive User Representation Learning

Inspired by the self-attention mechanism [25], we propose a socially-attentive representation learning method to incorporate social couplings into user representations. This method assumes users with strong social couplings should have similar representations. Accordingly, in the representation learning process, it adjusts the representation of each user by the representations of its neighbors in the space spanned by the coupling matrix.

Formally, the method adjusts the user representation matrix $\mathbf{V}_u$ as follows,

$$\mathbf{V}_u^* = \mathbf{C}^* \cdot \mathbf{V}_u, \tag{7}$$

where the weighting matrix $\mathbf{C}^*$ is calculated from the coupling matrix $\mathbf{C}$ that transforms each coupling value to a probability value, which is defined as follows,

$$\mathbf{C}^* = \begin{bmatrix} \frac{exp(C_{11})}{\sum_{i=1}^{n_u} exp(C_{1i})} & \frac{exp(C_{12})}{\sum_{i=1}^{n_u} exp(C_{1i})} & \cdots & \frac{exp(C_{1n_u})}{\sum_{i=1}^{n_u} exp(C_{1i})} \\ \frac{exp(C_{21})}{\sum_{i=1}^{n_u} exp(C_{2i})} & \frac{exp(C_{22})}{\sum_{i=1}^{n_u} exp(C_{2i})} & \cdots & \frac{exp(C_{2n_u})}{\sum_{i=1}^{n_u} exp(C_{2i})} \\ \vdots & \vdots & \ddots & \vdots \\ \frac{exp(C_{n_u 1})}{\sum_{i=1}^{n_u} exp(C_{n_u i})} & \frac{exp(C_{n_u 2})}{\sum_{i=1}^{n_u} exp(C_{n_u i})} & \cdots & \frac{exp(C_{n_u n_u})}{\sum_{i=1}^{n_u} exp(C_{n_u i})} \end{bmatrix}, \tag{8}$$

where $exp(\cdot)$ is the exponential function. Subsequently, it changes the representation learning objective function Eq. (1) as follows,

$$\begin{aligned} \min_{V,\Phi} \sum_{i=1}^{n_\nu} \sum_{y=\{0,1\}} & \mathbf{1}[y_i = y] \log q_i \\ + \sum_{\nu \in S} \sum_{\nu' \notin S} & \gamma \max\{0, 1 + \|\mathbf{v}_u^* + \mathbf{v}_o + \mathbf{v}_s - \mathbf{v}_r\|^2 \\ & - \|\mathbf{v}_{u'}^* + \mathbf{v}_{o'} + \mathbf{v}_{s'} - \mathbf{v}_{r'}\|^2\}, \end{aligned} \tag{9}$$

$$\begin{aligned} s.t. \ \ \gamma &= \begin{cases} 1 & u = u' \\ 0 & u \neq u', \end{cases} \\ q_i &= \mathrm{softmax}(\mathbf{w}D_\mathbf{p}([\mathbf{v}_u^*, \mathbf{v}_o, \mathbf{v}_s, \mathbf{v}_r]) + b), \quad <u, o, s, r> \in \nu_i, \\ \mathbf{v}_r &= t_\omega(r), \end{aligned}$$

where $\mathbf{v}_u^*$ is the adjusted representation of user u in $\mathbf{V}^*$. For a user, the weighting of its strongly coupled users will be large, while the weighting of the others will tend to be zero. As a result, the representation of each user will attend to the representations of its neighbors, and the representations of users with strong social couplings will be similar.

In Eq. (7), although the representation of a user is adjusted by its neighbors, the calculation involves the representations of all users. Consequently, it is time consuming in practical. To tackle this problem, for each user, SATURN truncates its top k nearest neighbors to adjust its representation. Accordingly, the learning objective of SATURN can be re-formalized as follows,

$$\min_{V,\Phi} \sum_{i=1}^{n_\nu} \sum_{y=\{0,1\}} \mathbf{1}[y_i = y] \log q_i$$
$$+ \sum_{\nu \in S} \sum_{\nu' \notin S} \gamma \max\{0, 1 + \|\mathbf{v}_u^* + \mathbf{v}_o + \mathbf{v}_s - \mathbf{v}_r\|^2$$
$$- \|\mathbf{v}_{u'}^* + \mathbf{v}_{o'} + \mathbf{v}_{s'} - \mathbf{v}_{r'}\|^2\},$$
$$s.t. \ \mathbf{v}_u^* = C_{uu}^* \mathbf{v}_u + \sum_{u^* \in N_k(u)} C_{uu^*}^* \mathbf{v}_{u^*},$$
$$\mathbf{v}_{u'}^* = C_{u'u'}^* \mathbf{v}_{u'} + \sum_{u'^* \in N_k(u')} C_{u'u'^*}^* \mathbf{v}_{u'^*},$$
$$\gamma = \begin{cases} 1 & u = u' \\ 0 & u \neq u', \end{cases}$$
$$q_i = \text{softmax}(\mathbf{w} D_\mathbf{p}([\mathbf{v}_u^*, \mathbf{v}_o, \mathbf{v}_s, \mathbf{v}_r]) + b), \quad <u,o,s,r> \in \nu_i,$$
$$\mathbf{v}_r = t_\omega(r),$$

(10)

where $N_k(u)$ refers to the set of k-nearest neighbors of u in the space spanned by the coupling matrix $\mathbf{C}$, and $C_{uu^*}^*$ is the entry value in $\mathbf{C}^*$ corresponding to user u and u^*.

3.5   Fraud Review Detection

SATURN detects fraud reviews according to its learned representations set V, networks with parameters Φ, social coupling matrix $\mathbf{C}$, and weighting matrix $\mathbf{C}^*$. Specifically, given a reviewing activity $\nu := <u,o,r,s>$, SATURN first looks up V for the representations $\mathbf{v}_u$, $\mathbf{v}_o$ and $\mathbf{v}_s$, and calculates review representation as $\mathbf{v}_r = t_\omega(r)$, where $\omega \in \Phi$ involves the parameters of SATURN's reviewing representation network $t_\omega(\cdot)$. Then, SATURN generates the socially-attentive user representation $\mathbf{v}_u^*$ as

$$\mathbf{v}_u^* = C_{uu}^* \mathbf{v}_u + \sum_{u^* \in N_k(u)} C_{uu^*}^* \mathbf{v}_{u^*}, \tag{11}$$

where $N_k(u)$ refers to the set of k-nearest neighbors of u in the space spanned by the coupling matrix $\mathbf{C}$, and $C_{uu^*}^*$ is the entry value in $\mathbf{C}^*$ corresponding to user u and u^*. In cold-start case, because $\mathbf{v}_u \notin V$, SATURN infers the socially-attentive representation of the new user according to the learned entities relations as follows,

$$\mathbf{v}_u^* = \mathbf{v}_r - \mathbf{v}_o - \mathbf{v}_s. \tag{12}$$

Then, SATURN identifies whether r is a fraud review via its learned classifier as follows,

$$q = \text{softmax}(\mathbf{w} D_\mathbf{p}([\mathbf{v}_u^*, \mathbf{v}_o, \mathbf{v}_s, \mathbf{v}_r]) + b), \tag{13}$$

where $\mathbf{w}, \mathbf{p}, b \in \Phi$ is the parameters of the classifier network.

4   Experiments

4.1   Data Sets

Following the literature [26,28] about cold-start fraud detection, our experiments are carried on four real-life data sets, including Yelp-hotel, Yelp-restaurant, Yelp-

NYC, and Yelp-Zip, which are also commonly used in previous fraud detection researches [19,21,24]. Table 1 displays the statistics of the data sets.

We split the original data sets into several subsets according to the time period to evaluate the fraud review detection performance in a stable way. We further split each subset into two parts by setting a time point. The first part includes the reviews posted before the time point, while the second part contains the rest reviews. From the second part, we pick up the reviews which are posted by new users at first time as cold-start reviews. We train the fraud detection methods on the first part and evaluate them on the second part.

Table 1. Data characteristics. In this table, #R refers to the number of reviews; #F and #FC refer to the number of fraud reviews and cold-start fraud reviews, respectively; and #N and #NC refer to the number of honest reviews and cold-start honest reviews, respectively.

Name	Training data		Testing data				
	Time period	#R	Time period	#F	#FC	#N	#NC
Zip_1	24/10/08 – 24/03/09	10,530	25/03/09 – 25/06/10	6,267	4,848	43,744	15,952
Zip_2	24/03/09 – 24/08/09	13,252	25/08/09 – 25/12/09	1,396	1,075	10,220	3,820
NYC_1	24/10/08 – 24/03/09	6,780	25/03/09 – 25/06/10	3,183	2,539	27,974	11,313
NYC_2	24/03/09 – 24/08/09	8,243	25/08/09 – 25/12/09	748	594	6,664	2,754

4.2   Evaluation Metrics

We evaluate the fraud review detection performance of each method by three metrics, including *precision*, *recall*, and *F-score*. Here, the precision evaluates the ratio of correct detected reviews over all detected reviews, recall reflects the ratio of undetected reviews over all relevant reviews, and the F-score indicates an average of precision and recall. We use all of them because the fraud detection in an imbalance classification problem [17], i.e. the number of fraud reviews are much less than honest reviews, that cannot be considered only from either precision or recall perspective. We report these three metrics per ground-truth honest and fraud classes to illustrate the performance for different categories, and further average them to show the overall performance. Higher precision, recall, and F-score indicate a better performance.

We follow the literature [24,26] to use the results of the Yelp commercial fake review filter as the ground-truth for performance evaluation. Although its filtered (fraud reviews) and unfiltered reviews (honest reviews) are likely to be the closest to real fraud and honest reviews [21], they are not absolutely accurate [11]. The inaccuracy exists because it is hard for the commercial filter to have the same psychological state of mind as that of the users of real fraud reviews who have real businesses to promote or to demote, especially for cold-start problems.

4.3   Parameters Settings

In our experiments, we use the pre-trained 100-dimensional word embedding by GloVe algorithm [23][2]. We embed the user, item and rating into a 100-dimension vector representation. We implement the fraud detector network by a 3-layer fully connected neural network with 100 nodes in the hidden layers and use ReLU as the activation function of all hidden nodes. We train our model by Adam [10] and batch size 32. For the parameters in the compared methods, we take their recommended settings.

4.4   Effectiveness on Cold-Start Fraud Detection

Experimental Settings. SATURN is compared with the state-of-the-art method JETB [26]. This method handling cold-start problem by considering entities (user, item and review) relations to embed reviews. When a new user posted a new review, this review can be represented by the trained network and classified by the classifier. In the literature [26], support vector machine (SVM) is used as the fraud classifier based on the JETB generated review features. However, SVM is with a time complexity $O(n^3)$, where n is the number of training samples. It is not suitable for the problem with large amount of data. To make JETB practicable, we use a 3-layer fully connected neural network instead of SVM as the fraud classifier of JETB.

We further compared with two review content-based fraud detection methods used in [26] as baseline competitors. Both of them extract features from review content, and feed these features into a classifier for fraud review detection. Specifically, the first method (denoted as Bigram) uses the bigram feature. The second method (denoted as Behavior) uses (i) the bigram feature, (ii) the length of review, (iii) the absolute rating diversity of a review compared with other reviews of the same item, and (iv) the similarity of a review to its most similar reviews of the same item under the cosine similarity. We also use 3-layer fully connected neural network as their fraud classifier.

Findings - SATURN Significantly Outperforming The State-of-the-art Cold-Start Fraud Detection Method. Table 2 illustrates the cold-start fraud detection performance of SATURN compared with JETB, Behavior, Bigram on four time period of Yelp-Zip and Yelp-NYC data sets. SATURN gains largely improvement for cold-start fraud review detection, i.e. 0.00, 0.08, 0.03, and 0.00 F-score increase on Zip_1, Zip_2, NYC_1, and NYC_2, respectively. This averaged performance improvement is mainly contributed by the increased recall for the fraud reviews (corresponding recall increase values are 0.02, 0.13, 0.05, and 0.03). As shown in the results, SATURN slightly "decreases" the performance of honest reviews detection. This "decreased" may be caused by the *noising ground-truth* of the cold-start fraud reviews that do not be detected by the Yelp commercial filter.

---

[2] The pre-trained word embedding can be downloaded from: http://nlp.stanford.edu/data/glove.6B.zip.

Table 2. Cold-start fraud detection performance of different methods. Precision (P), Recall (R) and F-score (F) are reported per normal and fraud reviews. The best results are highlighted in bold.

Data Info.		SATURN			JETB			Behavior			Bigram			Improvement		
Name	Category	P	R	F	P	R	F	P	R	F	P	R	F	P	R	F
Zip_1	Normal	0.77	0.88	0.82	0.77	**1.00**	**0.87**	0.77	0.99	**0.87**	**0.78**	0.96	0.86	−0.01	−0.12	−0.05
	Fraud	0.23	**0.12**	**0.16**	0.24	0.00	0.00	**0.54**	0.05	0.09	0.42	0.10	**0.16**	−0.31	**0.02**	**0.00**
Zip_2	Normal	**0.82**	0.84	0.83	0.78	**1.00**	**0.88**	0.79	0.99	**0.88**	0.80	0.92	0.85	**0.02**	−0.16	−0.05
	Fraud	0.33	**0.30**	**0.31**	0.45	0.01	0.02	**0.54**	0.06	0.11	0.37	0.17	0.23	−0.21	**0.13**	**0.08**
NYC_1	Normal	**0.83**	0.89	0.85	0.82	**1.00**	0.90	0.82	**1.00**	0.90	0.82	0.96	**0.89**	**0.01**	−0.11	−0.04
	Fraud	0.21	**0.13**	**0.16**	0.00	0.00	0.00	**0.38**	0.00	0.00	0.31	0.08	0.13	−0.17	**0.05**	**0.03**
NYC_2	Normal	0.82	0.85	0.84	0.82	**1.00**	**0.90**	0.82	**1.00**	**0.90**	**0.83**	0.94	0.88	−0.01	−0.15	**0.06**
	Fraud	0.18	**0.15**	**0.17**	0.00	0.00	0.00	0.00	0.00	0.00	**0.29**	0.12	**0.17**	**0.11**	**0.03**	**0.00**

In addition to review, SATURN further leverages information from user, item, and rating, which are guaranteed by the inferable representation and enable SATURN to effectively capture more fraud evidence from multiple views. As a result, SATURN can achieve significant performance improvement in cold-start fraud detection.

4.5 Effectiveness on General Fraud Detection

Experimental Settings. SATURN is further compared with JETB [26] and two state-of-the-art competitors: FRAUDER [7] and HoloScope [16] in detecting *general fraud reviews*, i.e. all the reviews contained in the testing data set. Different from JETB which is review content-based method, FRAUDER and HoloScopre are two social relation-based fraud review detection methods. Specifically, FRAUDER models the social relation as a graph and detects fraud reviews by dense subgraph mining. HoloScope also adopts graph to model social relation but detects fraud reviews by jointly considering the graph topology and review temporal spikes.

Findings - SATURN Significantly Improving General Fraud Detection Performance. The precision, recall and F-score of SATURN, JETB, FRAUDER, and HoloScope are reported in Table 3. Overall, SATURN significantly outperforms the competitors in fraud review detection. It improves 0.16, 0.21, 0.16, and 0.16 compared with the best-performing method in terms of F-score on four data sets for fraud review detection.

The dramatic performance improvement of SATURN is mainly contributed by jointly embedding user reviewing behavior and user/item social relations in its user-item-review-rating representations: (1) compared to FRAUDER and HoloScope that capture the social relations, SATURN further considers the user reviewing behavior to effectively detect personalized fraud; and (2) compared to JETB, SATURN seamlessly integrates user/item social relations to avoid camouflage. Consequently, SATURN obtains up to 0.25 recall improvement compared with the competitors.

Table 3. General fraud detection performance of different methods. Precision (P), Recall (R) and F-score (F) are reported per normal and fraud reviews. The best results are highlighted in bold.

Data Info.		SATURN			JETB			FRAUDER			HoloScope			Improvement		
Name	Category	P	R	F	P	R	F	P	R	F	P	R	F	P	R	F
Zip_1	Normal	**0.88**	0.90	0.89	0.87	**1.00**	**0.93**	0.87	0.95	0.91	0.86	0.86	0.86	0.01	−0.10	−0.04
	Fraud	**0.20**	**0.18**	**0.19**	0.18	0.00	0.01	0.01	0.00	0.00	0.03	0.03	0.03	0.02	0.15	0.16
Zip_2	Normal	**0.90**	0.87	0.88	0.78	**1.00**	0.88	0.88	0.95	**0.91**	0.87	0.88	0.88	0.02	−0.13	−0.03
	Fraud	0.22	**0.29**	**0.25**	**0.45**	0.01	0.02	0.04	0.02	0.02	0.04	0.04	0.04	−0.23	0.25	0.21
NYC_1	Normal	**0.91**	0.90	0.90	0.90	**1.00**	**0.95**	0.88	0.86	0.87	0.88	0.86	0.87	0.01	−0.10	−0.05
	Fraud	**0.16**	**0.17**	**0.17**	0.00	0.00	0.00	0.01	0.01	0.01	0.01	0.01	0.01	0.15	0.16	0.16
NYC_2	Normal	**0.91**	0.87	0.89	0.90	**1.00**	**0.95**	0.88	0.82	0.85	0.87	0.69	0.77	0.01	−0.13	−0.06
	Fraud	**0.16**	**0.22**	**0.19**	0.00	0.00	0.00	0.01	0.02	0.02	0.02	0.06	0.03	0.14	0.16	0.16

4.6 Evaluating the Effectiveness of Entities Relations and User Social Relations for Fraud Review Detection

Experimental Settings. We visualize the user representation in a two-dimensional space trough TSNE [18], and plot the ground-truth labels of each user at their positions in the representation space. The user representation learned according to entities relation embedding loss function Eq. (1) is compared with that learned according to socially-attentive representation learning loss function Eq. (10) on Yelp-Hotel and Yelp-Restaurant data sets.

Findings - Entities Relations-Embedded Representation Contributes to Personalized Fraud Review Detection and Social Relation-Embedded Representation Contributes To Collaborative Fraud Review Detection. The entities relations-embedded and social relations-embedded user representations are visualized in Fig. 2. It is shown users have more diverse representations in the entities relations-embedded representation space compared with social relations-embedded representation space. This indicates more personalized information is captured by the entities relations-embedded representation, which is important to identify personalized fraud reviews. However, in the entities relations-embedded representation space, the users with large density are not consistent with the ground-truth fraudster label. In contrast, the density of social relations-embedded representation is consistent with the ground-truth fraudsters distribution. As evidenced by [7], the collaborative manipulation of reviews will generate density connection between users. Accordingly, the results demonstrate our embedded social relation is essential for collaborative fraud review detection. A high quality user representation will enable a dense distribution for fraudsters because of the collaborative manipulation [7]. This qualitative illustrates that the social relations of users is essential for collaborative fraudsters detection.

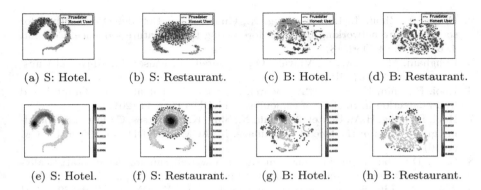

(a) S: Hotel.   (b) S: Restaurant.   (c) B: Hotel.   (d) B: Restaurant.

(e) S: Hotel.   (f) S: Restaurant.   (g) B: Hotel.   (h) B: Restaurant.

Fig. 2. User representation with density of different methods on Yelp-Hotel and Yelp-Restaurant. The sub-figures (a), (b), (c), (d) contain the user representation information with the ground-truth labels, and the sub-figures (e), (f), (g), (h) show the density in the representation space. S refers to the social relation embedding-based method, and B refers to the behavior embedding-based method.

5 Conclusion

This paper introduces a novel socially-attentive user representation method for fraud review detection with the cold-start problem. The proposed method jointly embeds the entities relations, the user social coupling relations, and the fraud-related information into a socially-attentive user representation space which provides sufficient evidence for fraud review detection. To comprehensively capture potential user social relations, it adopts a coupling learning method to learn both the explicit and implicit user social couplings. For the cold-start problem, the proposed method can effectively infer the socially-attentive representation of new users according to the learned entities relations. Four large real-world data sets demonstrate the performance of the proposed method significantly better than the state-of-the-art competitors.

Acknowledgement. This work was supported by the National Natural Science Foundation of China under Grant No. 61762033, and the National Natural Science Foundation of Hainan under Grant Nos. 2018CXTD333 and 617048.

References

1. Cheng, J., Li, M., Tang, X., Sheng, V.S., Liu, Y., Guo, W.: Flow correlation degree optimization driven random forest for detecting ddos attacks in cloud computing. Secur. Commun. Netw. **2018**, 1–14 (2018)
2. Cheng, J., Xu, R., Tang, X., Sheng, V.S., Cai, C.: An abnormal network flow feature sequence prediction approach for DDoS attacks detection in big data environment. CMC-Comput. Mater. Continua **55**(1), 095 (2018)
3. Cheng, J., Zhang, C., Tang, X., Sheng, V.S., Dong, Z., Li, J.: Adaptive DDoS attack detection method based on multiple-Kernel learning. Secur. Commun. Netw. **2018**, 1–19 (2018)

4. Cheng, J., Zhou, J., Liu, Q., Tang, X., Guo, Y.: A DDoS detection method for socially aware networking based on forecasting fusion feature sequence. Comput. J. **61**(7), 959–970 (2018)
5. Dehghani, M., Gouws, S., Vinyals, O., Uszkoreit, J., Kaiser, Ł.: Universal transformers. In: ICLR (2019)
6. Hooi, B., Shin, K., Song, H.A., Beutel, A., Shah, N., Faloutsos, C.: Graph-based fraud detection in the face of camouflage. TKDD **11**(4), 44 (2017)
7. Hooi, B., Song, H.A., Beutel, A., Shah, N., Shin, K., Faloutsos, C.: Fraudar: bounding graph fraud in the face of camouflage. In: ACM SIGKDD, pp. 895–904. ACM (2016)
8. Hovy, D.: The enemy in your own camp: how well can we detect statistically-generated fake reviews-an adversarial study. In: ACL, vol. 2, pp. 351–356 (2016)
9. Jindal, N., Liu, B.: Opinion spam and analysis. In: WSDM, pp. 219–230. ACM (2008)
10. Kingma, D.P., Ba, J.: Adam: a method for stochastic optimization. arXiv preprint arXiv:1412.6980 (2014)
11. Li, H., Chen, Z., Liu, B., Wei, X., Shao, J.: Spotting fake reviews via collective positive-unlabeled learning. In: ICDM, pp. 899–904. IEEE (2014)
12. Li, H., Chen, Z., Mukherjee, A., Liu, B., Shao, J.: Analyzing and detecting opinion spam on a large-scale dataset via temporal and spatial patterns. In: ICWSM, pp. 634–637 (2015)
13. Li, H., et al.: Modeling review spam using temporal patterns and co-bursting behaviors. arXiv preprint arXiv:1611.06625 (2016)
14. Li, Q., Wu, Q., Zhu, C., Zhang, J., Zhao, W.: Unsupervised user behavior representation for fraud review detection with cold-start problem. In: Yang, Q., Zhou, Z.-H., Gong, Z., Zhang, M.-L., Huang, S.-J. (eds.) PAKDD 2019. LNCS, vol. 11439, pp. 222–236. Springer, Cham (2019). https://doi.org/10.1007/978-3-030-16148-4_18
15. Liu, Q., Li, P., Zhao, W., Cai, W., Yu, S., Leung, V.C.M.: A survey on security threats and defensive techniques of machine learning: a data driven view. IEEE Access **6**, 12103–12117 (2018)
16. Liu, S., Hooi, B., Faloutsos, C.: Holoscope: topology-and-spike aware fraud detection. In: CIKM, pp. 1539–1548. ACM (2017)
17. Luca, M., Zervas, G.: Fake it till you make it: reputation, competition, and Yelp review fraud. Manag. Sci. **62**(12), 3412–3427 (2016)
18. van der Maaten, L., Hinton, G.: Visualizing data using t-SNE. JMLR **9**(11), 2579–2605 (2008)
19. Mukherjee, A., et al.: Spotting opinion spammers using behavioral footprints. In: ACM SIGKDD, pp. 632–640. ACM (2013)
20. Mukherjee, A., Venkataraman, V., Liu, B., Glance, N.: Fake review detection: classification and analysis of real and pseudo reviews. Technical report, Technical report UIC-CS-2013-03, University of Illinois at Chicago (2013)
21. Mukherjee, A., Venkataraman, V., Liu, B., Glance, N.S.: What yelp fake review filter might be doing? In: ICWSM (2013)
22. Ott, M., Choi, Y., Cardie, C., Hancock, J.T.: Finding deceptive opinion spam by any stretch of the imagination. In: ACL HLT, pp. 309–319. Association for Computational Linguistics (2011)
23. Pennington, J., Socher, R., Manning, C.: GloVe: global vectors for word representation. In: EMNLP, pp. 1532–1543 (2014)
24. Rayana, S., Akoglu, L.: Collective opinion spam detection: bridging review networks and metadata. In: ACM SIGKDD, pp. 985–994. ACM (2015)

25. Vaswani, A., et al.: Attention is all you need. In: NIPS, pp. 5998–6008 (2017)
26. Wang, X., Liu, K., Zhao, J.: Handling cold-start problem in review spam detection by jointly embedding texts and behaviors. In: ACL, vol. 1, pp. 366–376 (2017)
27. Ye, J., Akoglu, L.: Discovering opinion spammer groups by network footprints. In: Appice, A., Rodrigues, P.P., Santos Costa, V., Soares, C., Gama, J., Jorge, A. (eds.) ECML PKDD 2015. LNCS, vol. 9284, pp. 267–282. Springer, Cham (2015). https://doi.org/10.1007/978-3-319-23528-8_17
28. You, Z., Qian, T., Liu, B.: An attribute enhanced domain adaptive model for cold-start spam review detection. In: COLING, pp. 1884–1895 (2018)
29. Zhu, C., Cao, L., Liu, Q., Yin, J., Kumar, V.: Heterogeneous metric learning of categorical data with hierarchical couplings. TKDE **30**(7), 1254–1267 (2018)

Computational Model

Semi-online Machine Covering on Two Hierarchical Machines with Known Total Size of Low-Hierarchy Jobs

Man Xiao[1], Gangxiong Wu[1], and Weidong Li[1,2(✉)]

[1] School of Mathematics and Statistics, Yunnan University,
Kunming 650504, People's Republic of China
man1205@163.com, 1317630396@qq.com, weidongmath@126.com
[2] Dianchi College of Yunnan University,
Kunming 650000, People's Republic of China

Abstract. In this paper, we consider the online machine covering problem on two hierarchical machines with known the total size of low-hierarchy. We present several best possible online algorithms when some addition information is known or a buffer is given.

Keywords: Machine covering · Semi-online · Hierarchical machines · Competitive ratio

1 Introduction

In recent years, there have been extensive research on the hierarchical machine scheduling problem, which is to minimize the maximum machine load [7,10]. However, machine covering on hierarchical machines with the objective of maximizing the minimum machine load, denoted by $P|GoS|C_{min}$, is not a well-studied scheduling problem. For the offline case where information of all jobs is known in advance, Li et al. [8] presented a polynomial time approximation scheme for $P|GoS|C_{min}$.

Online scheduling over list is a scheduling where jobs arrive one by one, each job has to be scheduled with the current available information of arrived jobs. The performance of an online algorithm is measured by the competitive ratio. For a maximization scheduling problem and an online algorithm A, for any given instance I, let $C^A(I)$ be the objective value of the solution produced by A. The *competitive ratio* of the online algorithm A is defined as $\rho = \sup_I \frac{C^{OPT}(I)}{C^A(I)}$, where $C^{OPT}(I)$ is the optimal value of the instance I, i.e., $\rho C^A(I) \geq C^{OPT}(I)$ for any instance I. An online problem has a *lower bound* ρ if no online algorithm for the problem has a competitive ratio smaller than ρ. When an algorithm has the competitive ratio which is equal to the lower bound, it is called *the best possible*.

When the number of machines is two, $P|GoS|C_{min}$ is denoted by $P2|GoS|C_{min}$. Chassid and Epstein [1] first studied a generalized version of

© Springer Nature Singapore Pte Ltd. 2019
X. Sun et al. (Eds.): NCTCS 2019, CCIS 1069, pp. 95–108, 2019.
https://doi.org/10.1007/978-981-15-0105-0_7

$P2|GoS|C_{min}$ and showed that $P2|GoS|C_{min}$ has an unbounded competitive ratio. They also discussed a fraction online model and a semi-online model where the sum of jobs is known in advance. When we only know the size of the largest job, Wu et al. [11] showed that no algorithm with a bounded competitive ratio exists. They also designed two best possible algorithms for the case where we know the size of the largest job and its hierarchy. When the processing time of each job is bounded by an interval $[1, r]$, Luo et al. [9] presented a best possible online algorithm with a competitive ratio of $(1 + r)$ for $P2|GoS|C_{min}$. When the processing time of each job is in $\{1, 2, \ldots, 2^k\}$ where k is a positive integer, Wu and Li [12] presented a best possible online algorithm with a competitive ratio of 2^k. Epstein et al. [3] presented several best possible online algorithm for the machine covering with a reordering buffer.

In this paper, we consider the online machine covering problem on two hierarchical machines with known the total size of low-hierarchy. The rest of the paper is organized as follows. In Sect. 2, we present some basic notations. In Sect. 3, we give a best possible online algorithm when the total size of low-hierarchy is known. In Sect. 4, we give a best possible online algorithm when the total size of each hierarchy is known. In Sect. 5, we give a best possible online algorithm when the total size of low-hierarchy and the largest size are known. In Sect. 6, we give a best possible online algorithm when the total size of low-hierarchy is known and the processing time of each job is bounded by an interval $[1, r]$. In Sect. 7, we give a best possible online algorithm when the total size of low-hierarchy is known and a buffer is given. A summary and future research section concludes the paper.

2   Preliminaries

We are given two machines M_1, M_2, and a sequence of jobs arriving online which are to be scheduled irrevocably at the time of their arrivals. The first machine can process all the jobs while the second one can process only part of the jobs. The arrival of a new job occurs only after the current job is scheduled. Let $J = \{J_1, J_2, \ldots, J_n\}$ be the set of all jobs arranged in the order of arrival. We denote each job as J_j with p_j and g_j, where $p_j > 0$ is the *processing time* (also called job size) of the job J_j and $g_j \in \{1, 2\}$ is the hierarchy of the job J_j. If $J_j = 1$, job J_j must be processed by the first machine, and if $J_j = 2$, job J_j can be processed by either of the two machines. p_j and g_j are not known until the arrival of the job J_j. If $g_j = i$, job J_j is called the job of hierarchy i.

A schedule can be seen as the partition (S_1, S_2) of J, where S_i $(i = 1, 2)$ contains the jobs assigned to M_i. Let $L_i = \sum_{J_j \in S_i} p_j$ be the load of M_i, and T_i be the total processing time of the jobs of hierarchy i for $i = 1, 2$. The problem $P2|GoS|C_{min}$ is to find a schedule such that $\min\{L_1, L_2\}$ is maximized. For the first j jobs and $i = 1, 2$, let T_i^j be total processing time of the jobs of hierarchy i, and L_i^j be the total processing time of the jobs of hierarchy 2 scheduled on machine M_i after assigning job J_j. Clearly, $L_1 = L_1^n + T_1$, $L_2 = L_2^n$ and $T_i = T_i^n$ for $i = 1, 2$.

For convenience, let $UB^j = \min\{T_2^j, \frac{T_1^j + T_2^j}{2}, T_1^j + T_2^j - p_{max}^j\}$, for $j = 1, 2, \ldots, n$, where p_{max}^j is the largest processing time of the first j jobs. Then the following lemma can be got straightforwardly.

Lemma 1. *The optimal minimum machine load is at most UB^j after scheduling the job J_j for $j = 1, 2, \ldots, n$.*

Lemma 1 implies that

$$C^{OPT} \leq \min\{T_2, \frac{T_1 + T_2}{2}, T_1 + T_2 - p_{max}\}, \tag{1}$$

where $p_{max} = \max_j p_j$.

3 Known the Total Size of Low-Hierarchy

In this section, we consider the problem $P2|GoS, T_1|C_{min}$ where the total size of jobs of hierarchy 1 is given in advance. We give the lower bound 2 and prove that list-scheduling (LS) algorithm is best possible for this case.

Theorem 2. *Any on-line algorithm A for $P2|GoS, T_1|C_{min}$ has a competitive ratio at least 2.*

Proof. Assume that $T_1 = 1$. Consider the job sequence in [2]. For any algorithm A, the first two jobs in the sequence are $(1, 1)$ and $(1, 2)$, where the first number is the size and the second number is the hierarchy. If A assigns both of them to M_1, the sequence stops and $C^A = 0$, implying that the competitive ratio is unbounded. Else, if these two jobs are assigned to different machines, the last job is $(2, 2)$ and the sequence stops, then $C^{OPT} = 2$ and $\frac{C^{OPT}}{C^A} \geq 2$.

The main idea of LS proposed in [2] is to assign jobs with hierarchy of 2 to the machine with lower load (compare L_2^{j-1} with $L_1^{j-1} + T_1$, to get the machine with lower load). The details of LS is as follows.

Algorithm LS [2, 4]

Step 1. Let $j = 1$ and $L_1^{j-1} = L_2^{j-1} = 0$.

Step 2. If $g_j = 1$, schedule J_j on M_1 directly, and set $L_1^j = L_1^{j-1} + p_j$ and $L_2^j = L_2^{j-1}$.

Step 3. Else $g_j = 2$. If $L_1^{j-1} + T_1 \leq L_2^{j-1}$, schedule J_j on M_1, and compute L_1^j and L_2^j; else schedule J_j on M_2 and compute L_1^j and L_2^j.

Step 4. If there is another job, $j = j + 1$, go to **step 2**. Otherwise, stop.

Theorem 3. *The competitive ratio of ALGORITHM LS for $P2|GoS, T_1|C_{min}$ is 2, i.e. it is best possible.*

Proof. Let n be the number of the jobs, and L_i be the final load of machine M_i after assigning these n jobs. Clearly, $L_1 = T_1 + L_1^n$ and $L_2 = L_2^n$. If $L_1 = T_1$ and $L_2 = T_2 \leq T_1$, $C^{LS} = \min\{L_1, L_2\} = T_2 \geq C^{OPT}$, implying that it is optimal. Else, we distinguish two cases.

Case 1. $L_1 \leq L_2$. Let J_h be the last job assigned to M_2. According to ALGORITHM LS, we have $L_1 \geq L_2 - p_h$. Therefore,

$$C^{LS} = \min\{L_1, L_2\} = L_1 \geq \frac{L_1 + L_2 - p_h}{2} \geq \frac{T_1 + T_2 - p_{max}}{2} \geq \frac{1}{2}C^{OPT}.$$

Case 2. $L_1 > L_2$. As $L_1 > T_1$, there exists a job with hierarchy of 2 assigned to M_1. Let J_k be the last job assigned to M_1. According to ALGORITHM LS, we have $L_2 \geq L_1 - p_k$. Therefore,

$$C^{LS} = \min\{L_1, L_2\} = L_2 \geq \frac{L_1 + L_2 - p_k}{2} \geq \frac{T_1 + T_2 - p_{max}}{2} \geq \frac{1}{2}C^{OPT}.$$

Therefore, $\frac{C^{OPT}}{C^{LS}} \leq 2$ for any case.

4   Known the Total Size of Each Hierarchy

In this section, we assume that the total size of jobs of each hierarchy is given in advance, respectively, which means we know T_1 and T_2 in advance. This problem is denoted by $P2|GoS, T_1 \& T_2|C_{min}$. Similarly, we get the lower bound $\frac{3}{2}$ and prove that the algorithm common machine first (CMF) proposed in [2,5] has a competitive ratio of $\frac{3}{2}$.

Theorem 4. Any on-line algorithm A for $P2|GoS, T_1 \& T_2|C_{min}$ has a worst-case ratio at least $\frac{3}{2}$.

Proof. Consider the job sequence in [2] with $T_1 = 1$ and $T_2 = 5$. For any algorithm A, the first two jobs in the sequence are $(1, 1)$ and $(1, 2)$. If A assigns both of them to M_1, then the next two jobs are $(2, 2)$ and $(2, 2)$ and the sequence stops. We have $C^A = 2$ and $C^{OPT} = 3$, implying that $\frac{C^{OPT}}{C^A} \geq \frac{3}{2}$. If A assigns the first two jobs to different machines, the last two jobs are $(1, 2)$ and $(3, 2)$. We still have $C^A = 2$ and $C^{OPT} = 3$, implying that $\frac{C^{OPT}}{C^A} \geq \frac{3}{2}$.

The main idea of CMF proposed in [2,5] is to assign as many jobs with hierarchy of 2 as possible to M_1. The details of CMF is as follows.

Algorithm CMF [2, 5]

Step 1. If $T_2 \leq 2T_1$, schedule jobs with hierarchy of i to M_i directly ($i = 1, 2$), stop.

Step 2. Else $T_2 > 2T_1$, let $j = 1$ and $L_1^{j-1} = L_2^{j-1} = 0$.

    Step 2.1 If $g_j = 1$, schedule J_j on M_1 directly, and compute L_1^j and L_2^j.

    Step 2.2: Else $g_j = 2$. If $L_1^{j-1} + T_1 + p_j \leq (T_1 + T_2)/3$,

           schedule J_j on M_1, and compute L_1^j and L_2^j; go to **step 2.1** for the next job.

           Else let $t = j$, schedule J_t and the rest of jobs in the following way.

        Step 2.2.1: If $L_1^{t-1} + T_1 + p_t > 2(T_1 + T_2)/3$,

                schedule J_t on M_2 and the rest of jobs to M_1, stop.

        Step 2.2.2: If $L_1^{t-1} + T_1 + p_t \leq 2(T_1 + T_2)/3$,

                schedule J_t and the rest of jobs with hierarchy of 1 on M_1,

                and the rest of jobs with hierarchy of 2 to M_2, stop.

Theorem 5. The competitive ratio of ALGORITHM CMF for $P2|GoS,$ $T_1\&T_2|C_{min}$ is $\frac{3}{2}$, i.e. it is a best possible algorithm.

Proof. The proof is somewhat similar to that in [2]. We prove this theorem in detail for completeness. The algorithm can stop at **steps 1, 2.2.1** or **2.2.2**, so we need to consider the following three cases.

Case 1. ALGORITHM CMF stops at **step 1**. If $T_2 \leq T_1$, we have

$$C^{CMF} = \min\{T_1, T_2\} = T_2 \geq C^{OPT} \geq \frac{2}{3}C^{OPT},$$

where the first inequality follows from (1). If $T_1 < T_2 \leq 2T_1$, we have

$$C^{CMF} = \min\{T_1, T_2\} = T_1 \geq \frac{T_1 + T_2}{3} = \frac{2}{3}\frac{T_1 + T_2}{2} \geq \frac{2}{3}C^{OPT},$$

where the last inequality follows from (1). Thus, we have $\frac{C^{OPT}}{C^{CMF}} \leq \frac{3}{2}$ in any subcase.

Case 2. ALGORITHM CMF stops at **step 2.2.1**, i.e., $L_1^{t-1} + T_1 + p_t > 2(T_1 + T_2)/3$. Then, we have $L_1 = T_1 + T_2 - p_t$ and $L_2 = p_t > \frac{1}{3}(T_1 + T_2)$. If $L_1 < L_2$, we have

$$C^{CMF} = \min\{L_1, L_2\} = L_1 = T_1 + T_2 - p_t \geq T_1 + T_2 - p_{max} \geq C^{OPT},$$

where the last inequality follows from (1). Else if $L_1 \geq L_2$, we have

$$C^{CMF} = \min\{L_1, L_2\} = L_2 = p_t > \frac{1}{3}(T_1 + T_2) = \frac{2}{3}\frac{T_1 + T_2}{2} \geq \frac{2}{3}C^{OPT},$$

where the last inequality follows from (1). Thus, we have $\frac{C^{OPT}}{C^{CMF}} \leq \frac{3}{2}$ in any subcase.

Case 3. ALGORITHM CMF stops at **step 2.2.2**, i.e., $\frac{1}{3}(T_1 + T_2) < L_1^{t-1} + T_1 + p_t \leq \frac{2}{3}(T_1 + T_2)$. Then, we have $\frac{1}{3}(T_1 + T_2) < L_1 = L_1^{t-1} + T_1 + p_t \leq \frac{2}{3}(T_1 + T_2)$ and $\frac{1}{3}(T_1 + T_2) \leq L_2 = T_1 + T_2 - L_1^n < \frac{2}{3}(T_1 + T_2)$. Therefore, we have

$$C^{CMF} = \min\{L_1, L_2\} \geq \frac{1}{3}(T_1 + T_2) = \frac{2}{3}\frac{T_1 + T_2}{2} \geq \frac{2}{3}C^{OPT},$$

where the last inequality follows from (1).

Thus, we have $\frac{C^{OPT}}{C^{CMF}} \leq \frac{3}{2}$ in any case, i.e., the competitive ratio of ALGO-RITHM CMF for $P2|GoS, T_1\&T_2|C_{min}$ is $\frac{3}{2}$.

5 Known the Total Size of Low-Hierarchy and the Largest Size

In this section, we consider the problem $P2|GoS, T_1|C_{min}$ where the total size of jobs of hierarchy 1 and the largest processing time p_{max} of hierarchy 2 are given in advance. This problem is denoted by $P2|GoS, T_1\& \max|C_{min}$. We give the lower bound $\frac{3}{2}$ and propose an online algorithm with competitive ratio $\frac{3}{2}$. A job is called a *large* job if its processing time is p_{max}.

Theorem 6. Any on-line algorithm A for $P2|GoS,T_1\&\max|C_{min}$ has a worst-case ratio at least $\frac{3}{2}$.

Proof. Assume that $T_1 = 1$ and $p_{max} = 2$. The first two jobs are $J_1 = (1,1)$ and $J_2 = (1,2)$. If algorithm A assigns them to different machines, the last job $J_3 = (2,2)$ comes. Then $C^A = 1$ while $C^{OPT} = 2$. It follows that $\frac{C^{OPT}}{C^A} = 2$. If algorithm A assign the first two jobs to the same machine, then the incoming two jobs are $J_3 = (2,2)$ and $J_4 = (2,2)$. Then $C^A = 2$ while $C^{OPT} = 3$, implying that $\frac{C^{OPT}}{C^A} \geq \frac{3}{2}$.

Our algorithm is a modified version of the algorithm proposed in [6].

Algorithm MLS [6]

Step 1. Let $j = 1$ and $L_1^{j-1} = L_2^{j-1} = 0$.

Step 2. If $g_j = 1$, schedule J_j on M_1 directly, and compute L_1^j, L_2^j.

Step 3. Else $g_j = 2$.

    Step 3.1. If $L_2^{j-1} = 0$, do

        Step 3.1.1. If $p_j = p_{max}$, schedule J_j on M_2 and compute L_1^j, L_2^j;

        Step 3.1.2. If $p_j + L_1^{j-1} + T_1 > 2p_{max}$, schedule J_j on M_2 and compute L_1^j, L_2^j.

        Step 3.1.3. Else, schedule J_j on M_1, and compute L_1^j, L_2^j;

    Step 3.2. Else, $L_2^{j-1} > 0$, do

        Step 3.2.1. If $L_2^{j-1} \leq L_1^{j-1} + T_1$, schedule J_j on M_2 and compute L_1^j, L_2^j.

        Step 3.2.2. Else, schedule J_j on M_1, and compute L_1^j, L_2^j;

Step 4. If there is another job, $j = j + 1$, go to **step 2**. Otherwise, stop.

Theorem 7. The competitive ratio of ALGORITHM MLS for $P2|GoS$, $T_1\&max|C_{min}$ is $\frac{3}{2}$, i.e. it is a best possible algorithm.

Proof. Let J_h be the last job of hierarchy 2. We distinguish the following three cases.

    Case 1. $L_2^{h-1} = 0$.

It implies that $p_h = p_{max}$. ALGORITHM MLS will assign J_h to machine M_2 at **Step 3.1.1**. Thus, $L_1 = L_1^h + T_1 = L_1^{h-1} + T_1$ and $L_2 = L_2^h = p_h = p_{max}$. If $L_1 \leq p_{max}$, the schedule is optimal following from (1). Otherwise, we have $C^{MLS} = p_{max}$. By the choice of ALGORITHM MLS, we have $L_1^{h-1} + T_1 \leq 2p_{max}$. Hence,

$$C^{OPT} \leq \frac{L_1 + L_2}{2} = \frac{L_1^{h-1} + T_1 + p_{max}}{2} \leq \frac{3p_{max}}{2},$$

implying that $\frac{C^{OPT}}{C_{MLS}} \leq \frac{3}{2}$.

    Case 2. $0 < L_2^{h-1} \leq L_1^{h-1} + T_1$.

Job J_h will be assigned to M_2 at **Step 3.2.1** by ALGORITHM MLS. Thus, $L_1 = L_1^h + T_1 = L_1^{h-1} + T_1$, $L_2 = L_2^h = L_2^{h-1} + p_h$, and $C^{MLS} = \min\{L_1, L_2\} = \min\{L_1^{h-1} + T_1, L_2^{h-1} + p_h\}$.

    Case 2.1. $L_2^{h-1} < p_{max}$. Denote by J_k the first job assigned to M_2. By the choice of ALGORITHM MLS, we have

$$L_1^{k-1} + T_1 \leq 2p_{max}, \text{ and } p_k + L_1^{k-1} + T_1 > 2p_{max}.$$

Thus, $L_1^{h-1} + T_1 \geq L_1^{h-2} + T_1 \geq \cdots \geq L_1^{k-1} + T_1 > 2p_{\max} - p_k > p_{\max}$. Since $L_2^{k-1} \leq L_2^k \leq \cdots \leq L_2^{h-1} < p_{\max}$, for $j = k-1, k, \ldots, h-1$, we have $L_2^j < p_{\max} < L_1^j + T_1$, implying that all jobs $J_k, J_{k+1}, \ldots, J_{h-1}$ are not large jobs and assigned to M_2, i.e.,

$$p_{\max} < L_1^{h-1} + T_1 = L_1^{h-2} + T_1 = \cdots = L_1^{k-1} + T_1 \leq 2p_{\max}.$$

Therefore, J_h is a large job with $p_h = p_{\max}$, which implies that

$$p_{\max} < L_2^{h-1} + p_h \leq 2p_{\max}.$$

Following (1), we have

$$C^{OPT} \leq \frac{L_1^{h-1} + T_1 + L_2^{h-1} + p_h}{2} \leq \frac{L_1^{h-1} + T_1 + L_2^{h-1} + p_{\max}}{2}.$$

If $C^{MLS} = L_1 = L_1^{h-1} + T_1$, we have

$$\frac{C^{OPT}}{C^{MLS}} \leq \frac{L_1^{h-1} + T_1 + L_2^{h-1} + p_{\max}}{2(L_1^{h-1} + T_1)} \leq \frac{3(L_1^{h-1} + T_1)}{2(L_1^{h-1} + T_1)} = \frac{3}{2}.$$

If $C^{MLS} = L_2 = L_2^{h-1} + p_{\max}$, we also have

$$\frac{C^{OPT}}{C^{MLS}} \leq \frac{L_1^{h-1} + T_1 + L_2^{h-1} + p_{\max}}{2(L_2^{h-1} + p_{\max})} \leq \frac{2p_{\max} + L_2^{h-1} + p_{\max}}{2(L_2^{h-1} + p_{\max})} \leq \frac{3}{2}.$$

Case 2.2. $L_2^{h-1} \geq p_{\max}$. For this case, we have $L_1^{h-1} + T_1 \geq L_2^{h-1} \geq p_{\max}$. Therefore,

$$L_1^{h-1} + T_1 \geq \frac{1}{3}(L_1^{h-1} + T_1 + L_2^{h-1} + p_{\max}) \geq \frac{1}{3}(L_1^{h-1} + T_1 + L_2^{h-1} + p_h).$$

Since $L_2^{h-1} \geq p_{\max} > 0$, By the choice of ALGORITHM MLS at **Step 3.2**, we have $L_1^{h-1} + T_1 - L_2^{h-1} \leq p_{\max}$, i.e., $L_1^{h-1} + T_1 \leq L_2^{h-1} + p_{\max} \leq 2L_2^{h-1}$, implying that

$$L_2^{h-1} + p_h \geq \frac{1}{3}(2L_2^{h-1} + L_2^{h-1} + p_h) \geq \frac{1}{3}(L_1^{h-1} + T_1 + L_2^{h-1} + p_h).$$

Therefore,

$$\frac{C^{OPT}}{C^{MLS}} \leq \frac{L_1^{h-1} + T_1 + L_2^{h-1} + p_h}{2\min\{L_1^{h-1} + T_1, L_2^{h-1} + p_h\}} \leq \frac{3}{2}.$$

Case 3. $L_2^{h-1} > L_1^{h-1} + T_1$.

In this case, p_h will be assigned to M_1 at **Step 3.2.2**. By the choice of ALGORITHM MLS, we have $L_2^{h-1} \geq p_{\max}$. If $L_1^{h-1} + T_1 + p_h < p_{\max}$, by the choice of ALGORITHM MLS, M_2 accepts nothing but one large job with hierarchy of 2, implying that the schedule is optimal. Hence, we can assume that $L_1^{h-1} + T_1 + p_h \geq p_{\max}$. Now, we have

$$2L_2^{h-1} > L_1^{h-1} + T_1 + L_2^{h-1} \geq L_1^{h-1} + T_1 + p_{\max} \geq L_1^{h-1} + T_1 + p_h,$$

which implies

$$L_2^{h-1} > \frac{1}{3}(L_1^{h-1} + T_1 + L_2^{h-1} + p_h).$$

As before, since $L_2^{h-1} - (L_1^{h-1} + T_1) \leq p_{\max}$, we have

$$L_2^{h-1} \leq L_1^{h-1} + T_1 + p_{\max} \leq L_1^{h-1} + T_1 + L_1^{h-1} + T_1 + p_h \leq 2(L_1^{h-1} + T_1 + p_h),$$

which implies

$$L_1^{h-1} + T_1 + p_h = \frac{1}{3}(L_1^{h-1} + T_1 + p_h + 2(L_1^{h-1} + T_1 + p_h)) \geq \frac{1}{3}(L_1^{h-1} + T_1 + L_2^{h-1} + p_h).$$

Since $C^{MLS} = \min\{L_2^{h-1}, L_1^{h-1} + T_1 + p_h\}$, we have

$$\frac{C^{OPT}}{C^{MLS}} \leq \frac{L_1^{h-1} + T_1 + L_2^{h-1} + p_h}{2\min\{L_2^{h-1}, L_1^{h-1} + T_1 + p_h\}} \leq \frac{3}{2}.$$

From Theorem 6, we know the competitive ratio is best possible.

6   Known the Total Size of Low-Hierarchy and Tightly-Grouped Processing Times

In this section, we consider the problem $P2|GoS, T_1|C_{min}$ where the processing times of all jobs are in the interval $[1, r]$. This problem is denoted by $P2|GoS, T_1\&UB\&LB|C_{min}$. We give best possible competitive ratio for different values of r.

Theorem 8. If $r \geq 2$, the best possible competitive ratio for $P2|GoS, T_1\&UB\&LB|C_{min}$ is 2.

Proof. Assume that $T_1 = 1$. By Theorem 2, ALGORITHM LS has a competitive ratio 2. We will give an instance which shows that the ratio is tight. The first two jobs are $J_1 = (1, 1)$ and $J_2 = (1, 2)$. If any algorithm A assigns the first two jobs to M_1, and no further job comes any more, the ratio is unbounded. If algorithm A assigns them to different machines, then $J_3 = (r, 2)$ comes, and no more jobs come. Thus, $\frac{C^{OPT}}{C^A} = 2$ as desired.

Theorem 9. If $1 \leq r < 2$, the best possible competitive ratio for $P2|GoS, T_1\&UB\&LB|C_{min}$ is r.

Proof. Consider the following instances with $T_1 = 1$. The first two jobs are $J_1 = (1,1)$ and $J_2 = (1,2)$. If any algorithm A assigns the first two jobs to M_1, and no further job comes any more, the ratio is unbounded. If algorithm A assigns them to different machines, the next and last job $J_3 = (r,2)$ comes. Hence $C^A = 1$ while $C^{OPT} = r$, implying that $\frac{C^{OPT}}{C^A} \geq r$. Therefore, any online algorithm has a competitive ratio at least r.

Next, we will show ALGORITHM LS presented in Sect. 3 has a competitive ratio r. We will prove by contradiction that for any list of jobs.

If $\frac{C^{OPT}}{C^{LS}} > r$ and $L_1 = L_1^n + T_1 > L_2^n = L_2$, we have $C^{LS} = L_2^n \geq 1$ (otherwise, M_2 does not process any job). Since $L_1^n + T_1 + L_2^n \geq 2C^{OPT}$, we have

$$L_1^n + T_1 \geq 2C^{OPT} - C^{LS} > 2C^{OPT} - \frac{C^{OPT}}{r} = \frac{(2r-1)C^{OPT}}{r}.$$

Suppose J_h is the last job of hierarchy 2 scheduled on machine M_1. By the ALGORITHM LS, we have $L_1^{h-1} + T_1 \leq L_2^{h-1} \leq L_2^n = C^{LS}$. Thus,

$$p_h = L_1^n + T_1 - (L_1^{h-1} + T_1) > \frac{(2r-1)C^{OPT}}{r} - C^{LS} > \frac{(2r-2)C^{OPT}}{r}.$$

Suppose in the optimal schedule, the machine M_i which determines C^{OPT} has k jobs. Thus,

$$k \leq C^{OPT} \leq kr.$$

Since $C^{OPT} > rC^{LS} \geq r$, we have $k \geq 2$. By the fact $p_h \leq r < 2$, we have

$$2 > p_h > \frac{(2r-2)C^{OPT}}{r} \geq \frac{(2r-2)k}{r},$$

i.e., $(k-1)r < k$. Thus, the other machine M_{3-i} satisfies $L_{3-i} > C^{OPT} \geq k > (k-1)r$, implying that M_{3-i} has at least k jobs and there are at least $2k$ jobs.

Consider the solution produced by ALGORITHM LS. If there are at least k jobs scheduled on M_2, we have $C^{LS} \geq k$, which implies that $\frac{C^{OPT}}{C^{LS}} \leq r$. If there are less than k jobs scheduled on M_2 and $L_1^n = 0$, we have $C^{LS} = L_2^n = T_2 \geq C^{OPT}$, which implies LS algorithm produces an optimal solution. If there are less than k jobs scheduled on M_2 and $L_1^n > 0$, there are more than k jobs scheduled on M_1 before the last job J_t of hierarchy 2 scheduled on M_1. It implies that $L_1^{t-1} \geq k > (k-1)r \geq L_2^{t-1}$, contradicting the choice of ALGORITHM LS. Therefore, $\frac{C^{OPT}}{C^{LS}} \leq r$ must hold.

If $\frac{C^{OPT}}{C^{LS}} > r$ and $C^{LS} = L_1^n \geq 1$, we can prove the theorem similarly as above. $\qed$

7   Known Total Size of Low-Hierarchy and Given a Buffer

In this section, we consider the problem $P2|GoS, T_1|C_{min}$ where a buffer is available for storing at most one job. When the current job is given, we are allowed to assign it on some machine irrecoverably; or temporarily store it in the buffer. This problem is denoted by $P2|GoS, T_1\&buffer|C_{min}$. We give the lower bound $\frac{3}{2}$ and propose an online algorithm with competitive ratio $\frac{3}{2}$.

Theorem 10. Any on-line algorithm A for $P2|GoS, T_1\&buffer|C_{min}$ has a worst-case ratio at least $\frac{3}{2}$.

Proof. Assume $T_1 = 1$. The first jobs is $(1,1)$. All the next five jobs in the sequence are $(1,2)$. There are at most one job in the buffer. Without loss of generality, let $L_1^6 \le L_2^6$. If $L_1^6 \le 1$, the sequence stops. Then, $C^A \le L_1^6 + 1 \le 2$, and $C^{OPT} = 3$, implying that $C^{OPT}/C^A \ge \frac{3}{2}$. If $L_1^6 = 2$, the last job in the sequence is $(6,2)$. Then, $C^A \le 4$, and $C^{OPT} = 6$, implying that $C^{OPT}/C^A \ge \frac{3}{2}$.

Our algorithm is a modified version of the algorithm proposed in [3].

Algorithm BLS [3]

Step 1. Let $j = 1$ and $L_1^{j-1} = L_2^{j-1} = 0$.

Step 2. If $g_j = 1$, schedule J_j on M_1 directly, and compute L_i^j.

Step 3. Else $g_j = 2$. Consider the incoming job J_j and the job being in the buffer(if no job in the buffer, we assume there exist a job with size 0 in it). Put the bigger one in the buffer and let it be B_j (also using B_j to denote its size), and the smaller one be S_j (also denoting its size).

  Step 3.1. If $L_2^{j-1} + B_j \ge \frac{L_1^{j-1} + T_1 + S_j}{2}$, schedule S_j on M_1, and compute L_i^j.

  Step 3.2. Else, schedule S_j on M_2, and compute L_i^j.

Step 4. If there is another job, $j = j + 1$, go to step 2. Otherwise, schedule B_j on M_2, and stop.

Lemma 11. If job S_j is assigned to M_2, then

$$B_j < \frac{L_1^{j-1} + S_j + T_1}{2} - L_2^{j-1}, \text{ and } S_j \le L_1^{j-1} + T_1.$$

Proof. If S_j is assigned to M_2, we have $L_2^{j-1} + B_j < \frac{L_1^{j-1} + T_1 + S_j}{2}$, i.e., $B_j < \frac{L_1^{j-1} + T_1 + S_j}{2} - L_2^{j-1}$. If $S_j > L_1^{j-1} + T_1$, then

$$B_j < \frac{L_1^{j-1} + T_1 + S_j}{2} - L_2^{j-1} < S_j - L_2^{j-1} < S_j \le B_j.$$

A contraiction. Thus, the lemma holds.

Lemma 12. Let S_i be the first job of hierarchy 2 assigned to M_1. For all $j \ge i$, we have $L_2^{j-1} + B_j \ge \frac{L_1^{j-1} + T_1}{2}$.

Proof. It will be proved by induction. When $j = i$, by the definition of i, we have $L_1^{j-1} = 0$, and $L_2^{j-1} + B_j \ge \frac{L_1^{j-1} + T_1 + S_j}{2} > \frac{L_1^{j-1} + T_1}{2}$, implying that the lemma holds when $j = i$. Suppose that the lemma holds before S_j is assigned.

If S_j is assigned M_1, we have $L_1^j = L_1^{j-1} + S_j$, and $L_2^j = L_2^{j-1}$. By the algorithm, we have

$$L_2^{j-1} + B_j \ge \frac{L_1^{j-1} + T_1 + S_j}{2}.$$

Since $B_{j+1} \geq B_j$, then

$$L_2^j + B_{j+1} \geq L_2^{j-1} + B_j \geq \frac{L_1^{j-1} + T_1 + S_j}{2} \geq \frac{L_1^j + T_1}{2}.$$

If S_j is assigned M_2, we have $L_1^j = L_1^{j-1}$, and $L_2^j > L_2^{j-1}$. Since $B_{j+1} \geq B_j$, then

$$L_2^j + B_{j+1} \geq L_2^{j-1} + B_j \geq \frac{L_1^{j-1} + T_1}{2} = \frac{L_1^j + T_1}{2}.$$

Hence, for any case, $L_2^j + B_{j+1} \geq \frac{L_1^j + T_1}{2}$, implying that the claim holds.

Lemma 13. If $L_1^{j-1} + T_1 < 2L_2^{j-1}$, and the last job assigned to M_2 is S_i $(i < j)$, then the job B_j in the buffer is exactly B_i, and $L_1^{j-1} + T_1 + S_i > 2L_2^{j-1}$.

Proof. By the definition of i, we have

$$L_2^{j-1} = L_2^i = L_2^{i-1} + S_i \leq L_2^{i-1} + B_i.$$

By the algorithm,

$$L_2^{i-1} + B_i < \frac{L_1^{i-1} + T_1 + S_i}{2} \leq \frac{L_1^{j-1} + T_1 + S_i}{2},$$

implying that $L_1^{j-1} + T_1 + S_i > 2L_2^{j-1}$. By the assumption $L_1^{j-1} + T_1 < 2L_2^{j-1}$, we have

$$L_1^{i-1} + T_1 + S_i > 2(L_2^{i-1} + B_i) \geq 2L_2^{j-1} > L_1^{j-1} + T_1,$$

implying that $L_1^{j-1} - L_1^{i-1} < S_i \leq B_i$, i.e., the total size of jobs with hierarchy of 2 between J_i and J_{j-1} is strictly lower than B_i. This shows that job B_i is still in the buffer after S_{j-1} is assigned.

Theorem 14. The competitive ratio of this algorithm is at most $\frac{3}{2}$.

Proof. Let J_n be the last job with hierarchy of 2. Following (1), we have

$$C^{OPT} \leq \frac{L_1^{n-1} + T_1 + L_2^{n-1} + S_n + B_n}{2}.$$

Case 1. S_n is assigned to M_1. By the algorithm, we have

$$2(L_2^{n-1} + B_n) \geq L_1^{n-1} + T_1 + S_n.$$

Case 1.1. $L_2^{n-1} + B_n \leq L_1^{n-1} + T_1 + S_n$. Then, $C^{BLS} = L_2^{n-1} + B_n$, and the competitive ratio is

$$\frac{C^{OPT}}{C^{BLS}} \leq \frac{L_1^{n-1} + T_1 + L_2^{n-1} + S_n + B_n}{2(L_2^{n-1} + B_n)} = \frac{1}{2} + \frac{L_1^{n-1} + T_1 + S_n}{2(L_2^{n-1} + B_n)} \leq \frac{3}{2}.$$

Case 1.2. $L_1^{n-1} + T_1 + S_n < L_2^{n-1} + B_n \le 2(L_1^{n-1} + T_1 + S_n)$. Then, $C^{BLS} = L_1^{n-1} + T_1 + S_n$, and the competitive ratio is

$$\frac{C^{OPT}}{C^{BLS}} \le \frac{L_1^{n-1} + T_1 + L_2^{n-1} + S_n + B_n}{2(L_1^{n-1} + T_1 + S_n)} = \frac{1}{2} + \frac{L_2^{n-1} + B_n}{2(L_1^{n-1} + T_1 + S_n)} \le \frac{3}{2}.$$

Case 1.3. $L_2^{n-1} + B_n > 2(L_1^{n-1} + T_1 + S_n)$ and $L_1^{n-1} + T_1 + S_n \ge 2L_2^{n-1}$. Then $C^{BLS} = L_1^{n-1} + T_1 + S_n$, and

$$B_n > 2(L_1^{n-1} + T_1 + S_n) - L_2^{n-1} \ge L_1^{n-1} + T_1 + L_2^{n-1} + S_n,$$

implying $C^{OPT} = L_1^{n-1} + T_1 + L_2^{n-1} + S_n$. Hence,

$$\frac{C^{OPT}}{C^{BLS}} = \frac{L_1^{n-1} + T_1 + L_2^{n-1} + S_n}{L_1^{n-1} + T_1 + S_n} = 1 + \frac{L_2^{n-1}}{L_1^{n-1} + T_1 + S_n} \le \frac{3}{2}.$$

Case 1.4. $L_2^{n-1} + B_n > 2(L_1^{n-1} + T_1 + S_n)$ and $L_1^{n-1} + T_1 + S_n < 2L_2^{n-1}$. Then, $L_1^n = L_1^{n-1} + S_n$, and $L_2^n = L_2^{n-1}$. Let S_i be the last job assigned to M_2. If $S_n = B_i$, by Lemma 13, we have

$$L_1^{n-1} + T_1 + S_n = L_1^{n-1} + T_1 + B_i \ge L_1^{n-1} + T_1 + S_i > 2L_2^{n-1},$$

which leads to a contradiction. Therefore $B_n = B_i$. Therefore, by the algorithm,

$$L_2^{i-1} + B_n = L_2^{i-1} + B_i < \frac{1}{2}(L_1^{i-1} + T_1 + S_i),$$

and on the other hand

$$L_2^{i-1} + S_i + B_n = L_2^{n-1} + B_n > 2(L_1^{n-1} + T_1 + S_n).$$

Therefore,

$$S_i > 2(L_1^{n-1} + T_1 + S_n) - (L_2^{i-1} + B_n) > 2(L_1^{n-1} + T_1 + S_n) - \frac{1}{2}(L_1^{i-1} + T_1 + S_i),$$

i.e.,

$$\begin{aligned} S_i &> \frac{2}{3}\left(2(L_1^{n-1} + T_1 + S_n) - \frac{1}{2}(L_1^{i-1} + T_1)\right) \\ &= \frac{4}{3}(L_1^{n-1} + T_1 + S_n) - \frac{1}{3}(L_1^{i-1} + T_1) \\ &= L_1^{n-1} + \frac{1}{3}(L_1^{n-1} - L_1^{i-1}) + T_1 + S_n \\ &> L_1^{n-1} + T_1, \end{aligned}$$

which contradicts the result from Lemma 11.

Case 2. The job S_n is assigned to M_2.

If after the assignment of B_n, M_1 has the lowest load, then the objective value of the returned solution is $L_1^{n-1} + T_1$, and the optimal value is at most

$$\frac{1}{2}(L_1^{n-1} + T_1 + L_2^{n-1} + S_n + B_n) \leq \frac{1}{2}(L_1^{n-1} + T_1 + L_2^{n-1} + S_n + \frac{L_1^{n-1} + T_1 + S_n}{2} - L_2^{n-1})$$

$$= \frac{3}{4}(L_1^{n-1} + T_1 + S_n)$$

$$\leq \frac{3(L_1^{n-1} + T_1)}{2},$$

where the inequalities follows from Lemma 11. Therefore, $\frac{C^{OPT}}{C^{BLS}} \leq \frac{3}{2}$.

If M_1 has the highest load, the profit of the algorithm is $C^{BLS} = L_2^{n-1} + S_n + B_n$. By Lemma 12, the competitive ratio is

$$\frac{C^{OPT}}{C^{BLS}} \leq \frac{L_1^{n-1} + T_1 + L_2^{n-1} + S_n + B_n}{2(L_2^{n-1} + S_n + B_n)}$$

$$= \frac{1}{2} + \frac{L_1^{n-1} + T_1}{2(L_2^{n-1} + S_n + B_n)}$$

$$\leq \frac{1}{2} + \frac{2(L_2^{n-1} + B_n)}{2(L_2^{n-1} + S_n + B_n)}$$

$$\leq \frac{3}{2}.$$

This completes the proof.

8   Conclusion

We have presented several best possible online algorithms for the online machine covering problem on two hierarchical machines with known the total size of low-hierarchy, when some addition information is known or a buffer is given. It is an interesting problem to design online algorithms for the case where there are more than two hierarchical machines.

Acknowledgements. The work is supported in part by the National Natural Science Foundation of China [No. 61662088], IRTSTYN, and Key Joint Project of the Science and Technology Department of Yunnan Province and Yunnan University [No. 2018FY001(-014)].

References

1. Chassid, O., Epstein, L.: The hierarchical model for load balancing on two machines. J. Comb. Optim. **15**(4), 305–314 (2008)
2. Chen, X., Ding, N., Dósa, G., Han, X., Jiang, H.: Online hierarchical scheduling on two machines with known total size of low-hierarchy jobs. Int. J. Comput. Math. **92**(5), 873–881 (2015)

3. Epstein, L., Levin, A., van Stee, R.: Max-min online allocations with a reordering buffer. SIAM J. Discrete Math. **25**(3), 1230–1250 (2011)
4. Graham, R.L.: Bounds for certain multiprocessing anomalies. Bell Syst. Tech. J. **45**(9), 1563–1581 (1966)
5. Kellerer, H., Kotov, V., Speranza, M.G., Tuza, Z.: Semi on-line algorithms for the partition problem. Oper. Res. Lett. **21**, 235–242 (1997)
6. He, Y.: Semi-on-line scheduling problems for maximizing the minimum machine completion time. Acta Mathematicae Applicatae Sinica (Engl. Ser.) **17**(1), 107–113 (2001)
7. Leung, J.Y.T., Li, C.L.: Scheduling with processing set restrictions: a literature update. Int. J. Prod. Econ. **175**, 1–11 (2016)
8. Li, J., Li, W., Li, J.: Polynomial approximation schemes for the max-min allocation problem under a grade of service provision. Discrete Math. Algorithms Appl. **1**(3), 355–368 (2009)
9. Luo, T., Xu, Y.: Semi-online hierarchical load balancing problem with bounded processing times. Theoret. Comput. Sci. **607**, 75–82 (2015)
10. Tan, Z., Zhang, A.: Online and semi-online scheduling. In: Pardalos, P.M., Du, D.-Z., Graham, R.L. (eds.) Handbook of Combinatorial Optimization, pp. 2191–2252. Springer, New York (2013). https://doi.org/10.1007/978-1-4419-7997-1_2
11. Wu, Y., Cheng, T.C.E., Ji, M.: Optimal algorithm for semi-online machine covering on two hierarchical machines. Theoret. Comput. Sci. **531**, 37–46 (2014)
12. Wu, G., Li, W.: Semi-online machine covering on two hierarchical machines with discrete processing times. Commun. Comput. Inf. Sci. **882**, 1–7 (2018)

A Combined Weighted Concept Lattice for Multi-source Semantic Interoperability of ECG-Ontologies Based on Inclusion Degree and Information Entropy

Kai Wang[✉] 🆔

BengBu Medical College, BengBu 233000, Anhui, China
Wangkai0552@126.com

Abstract. To deal with the complexities associated with the rapid growth in a merged concept lattice, a formal method based on an entropy-based weighted concept lattice is proposed. This paper solves the problem by constructing the fundamental relation of partial order generated from the weighted formal context to get a weighted concept lattice. The results indicate that the proposed method is feasible and valid for reducing the complexities associated with the merging of ECG-ontologies.

Keywords: Weighted concept lattice · Entropy · Inclusion degree

1 Introduction

Biomedical data has been ever more available for computing as the use of information systems in Biology and Medicine becomes widespread. An example of biomedical data is the electrocardiogram (ECG), which is the most applied test for measuring heart activity in Cardiology. Among the prominent ECG standards, one might refer to different objects of standardization initiatives, such as AHA/MIT-BIH (Physionet) [1], SCP-ECG [2] and HL7 aECG [3]. However, in spite of the fact that all of them address the very same ECG domain, the conceptualizations underlying these standards are heterogeneous, due to the principles that are mostly on how data and information represented in computer and messaging systems [4]. One way to address this challenge is with ECG reference ontologies (henceforth, just "ECG-ontologies") integration, which is a formal solution to integrate heterogeneous ECG waveform data from multiple sources. For ECG reference ontologies [5], we mean ontologies resulting from an application-independent representation of a given domain to address the heterogeneity issues in an effective manner and achieve semantic integration between ECG data standards. The ECG-ontologies used in this article has been carefully developed over the years by employing ontological principles coming from the discipline of formal ontology in philosophy [6]. Thus, in this paper, just as mentioned, integration of ECG-ontologies is the procedure of solving the problem of heterogeneous ECG data formats.

© Springer Nature Singapore Pte Ltd. 2019
X. Sun et al. (Eds.): NCTCS 2019, CCIS 1069, pp. 109–130, 2019.
https://doi.org/10.1007/978-981-15-0105-0_8

Up to now, there have been some significant methods of integrating ontologies proposed in recent decades, among which formal concept analysis (FCA) is a key progress in merging domain ontologies in this research field. Abdelkader [7] presented an approach to locate services in legacy software, by means of information retrieval techniques using the FCA techniques to reduce the search space time. Similarly, Fu [8] proposed a formal and semi-automated approach for ontology development based on FCA to integrate data that exhibits implicit and ambiguous information. In the field of biomedical information science, using the same principle, Fang [9] explored integrating ontology-based Traditional Chinese Medicine (TCM) by constructing a unified concept lattice, the aim of which is to contribute towards eliminating the phenomenon of information disparity between patients and physicians. Yang [10] introduced a method into medical fields by using ontology technique to construct a medical term system on the basis of FCA tools. The visualized display of FCA-based concept lattice can reflect the hierarchical relationship of domain terms clearly. In addition, Kim [11] explored the potential role of FCA in augmenting an existing ontology in medical domain by applying natural language processing techniques and presented the details of MeSH ontology augmentation.

By integrating ECG-ontologies as a reference to meet Cimino's desiderata [12], the elements present in these data formats could be semantically mirrored to the ontology entities. Thereby, extracting essential ontological properties to build a formal context, instead of being objects of pairwise mappings, and removing redundant concepts and relations to reduce the concept lattice are the two key tasks, regardless of technological issues that arise in representing it in a given information system. Then it could be used to extract the formal representation of the complexity in different levels, due to the rapid growth of the merged concept lattice. However, the previous methodology applied have drawbacks in two aspects, namely: (i) the methods above mainly get attention to refine properties of non-vagueness, lack of taking different valued attributes into account, always assuming that they are equally important in the process of constructing cardiac physiology research [13]. (ii) integration methods conducting a concept reduction always involve numerous randomness and ambiguity, to which the ECG-ontologies axiomatization allows little freedom, forcing the data formats to make their assumptions explicit. Besides, the proposal of handling these shortcomings is also cost-effective both in space and time complexity, especially for constructing a concept lattice with dense and large contexts.

Following this premise, we can observe that the contribution of this paper is as follows:

(1) Elicit the conceptualizations underlying the ECG standards aforementioned to explicitly represent them as conceptual models;
(2) Introduce a method for computing the weight of given formal concepts using the inclusion degree and Shannon entropy;
(3) Provide the link between rough set theory and concept lattice for simplifying the size of concept lattice using the different thresholds of semantic granularity.

The remainder of this article is organized as follows: Sect. 2 analyses the conceptual model from semantic specifications based on the textual descriptions and provides a brief background about formal concept analysis which is germane to the purposes of this article; Sect. 3 describes the basic notions of the illustration of rough set theory and defines attribute importance based on the inclusion degree; Sect. 4 discusses the materials and methods of calculating the weighted attributes and simplifying the weighted concept lattice structure from the point of view of attribute reduction; Sect. 5 gives the semantic link between different levels of conceptual granularity and weighted concept lattice for supporting an in-depth ontological reading. Finally, Sect. 6 contributes and outlines the direction of the future work.

2 A Semantic Representation of ECG-Ontologies Concepts with Formal Concept Analysis

2.1 Concepts of ECG-Ontologies

Every design or implementation artifact commits to an underlying conceptualization. However, this commitment is made in an implicit frequently. Among the prominent ECG standards, one might refer to (i) AHA/MIT-BIH [14], whose data format is extensively used worldwide in cardiac physiology research; (ii) SCP-ECG [2], which is an ECG European standard that specifies a data format and a transmission procedure for ECG records; and (iii) HL7 aECG [3], which is an ECG data standard adopted by FDA for clinical trials [15]. Given the various existence of ECG standards, we put in focus the objects of ECG analysis, which are most relevant to the presentation of the ECG-ontologies listed in Table 1 alongside the corresponding definitions.

Limited to the length of this article, we mainly concentrate on excavating the conceptualizations underlying the ECG data standards based on the ECG-ontologies referenced from the standards of AHA/MIT-BIH and SCP-ECG, to achieve the conceptual descriptions of their composing properties. Meanwhile, we choose the beats marked by their QRS complex with a label called an annotation in BIH DB to eliminate the semantic heterogeneity between different ontologies. Afterwards, a feature of more signals is described at a given time instant in the record, and the feature to be annotated is the type of the beat (normal, ventricular ectopic, etc.). Correspondingly, in order to facilitate the comparison of ECG fields in BIH, we refer here only to some of the most important SCP elements in the SCP data standard. Through the analysis of SCP-ECG elements in Sect. 2.2, we extracted the record interpretation made further by physicians, such as reference beat and rhythm data. The conceptual model resulting from our excavation of these two standards is depicted in the Fig. 1, using standard UML diagrams, which is a de facto standard for creating visual conceptual models in computer science. For further introduction to UML, the reader should refer to [16].

Table 1. Entities of the ECG-ontologies that is relevant for the semantic integration of ECG standards.

Term	Entity ID	Textual definition
Record	ecgOnto:085	ECG data record (in any medium) resulting from a recording session and essentially composed of an ECG Waveform
Waveform	ecgOnto:091	Non-elementary Geometric form constituted by the sample sequence resulting from the observation series which makes up an ECG recording session
Observation	ecgOnto:092	Measurement of the p.d. between two regions of the patient's body. It is carried out by an ECG Recording device by means of two Electrode placements on those regions. The placements are defined according to an ECG Lead
Sample sequence	ecgOnto:095	Ordered sequence of samples resulting from an observation series
Lead	ecgOnto:096	Viewpoint of the heart activity that emerges from an observation series of the p.d. between two electrode placements on specific regions of the surface of the patient's body
Cycle	ecgOnto:105	Elementary form periodically repeated in the ECG waveform that indirectly indicates a heartbeat. The QRS complex is an essential part of it, as the peak of the R wave is considered a reference point to define it

We use here the term "conceptual model excavation" to refer to the activity of making explicit these implicit conceptualizations, followed by their representations via concrete engineering artifacts. In order to enlarge the application of this paper, we mainly discuss the waveforms resulting from the process of ECG acquisition as well as the interpretation to provide a tracing mechanism. Heart beats are mirrored to cardiac cycles that compose the ECG waveform, which are the forms appearing within every cardiac cycle that directly maps a cohesive and electrophysiological event in the heart behavior. For instance, the P wave maps the depolarization of atria, while the QRS complex maps the depolarization of ventricles and the PR segment, rather, connects the former to the latter. From the given normative descriptions, heterogeneity problems inevitably exist. For example, one considers a kind of disease named "*bundle branch block*" which consists of the following semantic properties and relations: waveform with value "*rhythm data*", segment morphology with value "*ST depressions and elevations*", waves with the value "*QRS Complex wide and T wave inversion*", data type of the period with value "*sample time interval*" and the other defines the same disease associated with the corresponding semantic properties: waveform with the value "*signal*", spatial morphology with the value "*ST depressions and elevations*", subkind function with the value "*baseline*", datatype of sample rate (Hz) with the value "*sampling frequency* (Hz)". In terms of the formal definitions of ontologies, ECG-ontologies are applied to capture the universal concepts and meanings in the biological signal domain. However, "*Acute Myocardial Infarction (AMI)*: *The elevation of Q wave and ST segments with the dynamic evolution of ST-T segments*" has a bit strong subjectivity and a non-rigid property, because different patients may have different clinical symptoms. In the acute phase of *AMI*, as the *ST* segment of the *AVF* lead is used for oblique elevation and the *Q* wave has a shorter duration, the clinical features of *AMI* is not easily recorded, which is often misdiagnosed as some irregular *ST-T* dynamic evolution. Hence, to improve the diagnostics effects based on machine learning, six typical cases with reference beats in StPetersbug INCHART 12-lead Arrhythmia Database [17] are selected to be used as objects and the symptoms of diseases related to these cases as sets to attributes. The result of partial semantics of concepts with reference beats is shown as Table 2.

Table 2. Partial semantics of concepts between entities in the ECG-ontologies and corresponding standards.

Object	Normative description	Ontological properties
AMI	Tissue death (infarction) of the heart muscle (myocardium), a type of acute coronary syndrome, which describes a sudden or short-term change in symptoms related to blood flow to the heart	Waveform/(T-inversion, Q-Duration), Segment/ST-depression, Reference beat/QRS, Complex/widening
CAD	Disease refers to a group of diseases which includes stable angina, unstable angina, myocardial infarction, and sudden cardiac death	Waveform/T-AmplitudeRatio Segment/ ST-depression, Reference beat/QSP, Complex/Duration
HYP	An elevated level of potassium (K+) in the blood serum	Waveform/P-loss, Reference beat/QRS, Complex/Widening
AF	An abnormal heart rhythm characterized by rapid and irregular beating of the atria, which starts as brief periods of abnormal beating and becomes longer and possibly constant over time	Waveform/P-loss, Reference beat/QRS, Segment/RR-inordinance, Complex/Duration
BBB	The conduction system abnormalities below the atrioventricular node can start from various diseases	Waveform/(R-Amplitude, T-AmplitudeRatio), Lead/(I, V1), Reference beat/QRS, Complex/Duration
WS	An electrocardiographic manifestation of critical proximal left anterior descending (LAD) coronary artery stenosis in patients with unstable angina. It is characterized by symmetrical, often deep (>2 mm)	Waveform/(T-inversion), Lead/(V2, V3), Reference beat/ST, Segment/no-ST- elevation

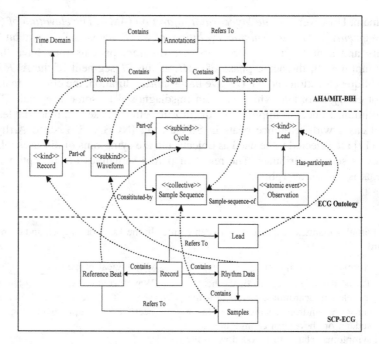

Fig. 1. Integration between two conceptual models of ECG data standards.

2.2   Concepts from the View of Formal Concept Analysis

Concepts and Concept Lattices. Formal concept analysis (FCA), proposed by Wille in 1982, is a branch of applied mathematics based on lattice theory, which has been applied to various fields, such as medical knowledge discovery, information science, and computer science [18, 19]. To analyze and visualize the relations between objects and attributes in a given application domain, a conceptual framework of hierarchies is built which is defined as a formal context consisting of an object set and an attribute set. Within the formal context, the whole concepts could be constructed into a concept lattice using partial order relation. In what follows, we briefly describe some basic analysis by FCA. A more detailed analysis of FCA can be found in [20].

Definition 1. (Many-valued context) A many-valued context is described by a triple $K_m = (O, P, W, R)$, where O and P are two nonempty sets of objects and attributes, respectively, and W is called attribute values, representing a subset of the Cartesian product of O and P ($W \subseteq O \times P$). R is a ternary relation, i.e., ($R \subseteq O \times P \times W$) between those three sets for which it holds that $(o, p, w) \in R$. When $o \in O$ and $p \in P$, if oRp, it means that object o has the attribute p, and that attribute p belongs to object o.

In a many-valued context $K_m = (O, P, W, R)$, for two given sets, $A \subseteq O$ and $B \subseteq P$, all attributes $p \in P$ are shared by objects from A. Similarly, all objects $o \in O$ are the attributes from B in common. A formal concept of K_m is derived in terms of the following operations:

$$A = B^{\downarrow} = \{a \in O | b \in B, aRb\} \tag{1}$$

$$B = A^{\uparrow} = \{b \in P | a \in A, aRb\} \tag{2}$$

Then, the formal concept is a pair of (A, B) satisfying $A^{\uparrow} = B$ and $B^{\downarrow} = A$, where the set of objects A called as the extent and the set of attributes B called as the intent. For the pair $(\uparrow, \downarrow)$ is known as a Galois connection [20]. The two sets of formal concepts $((A_1, B_1), (A_2, B_2))$ generated from a given formal context K_m define the partial ordering principle on condition that they satisfy the following formula:

$$(A_1, B_1) \leq (A_2, B_2) \Leftrightarrow A_1 \subseteq A_2 \Leftrightarrow B_2 \subseteq B_1 \tag{3}$$

where " $\leq$ " in the above condition represents the hierarchical relation between (A_1, B_1) and (A_2, B_2). What's more, (A_1, B_1) is called a sub-concept of (A_2, B_2) while (A_2, B_2) is called a super-concept of (A_1, B_1). If the set of all concepts in K_m is indicated as $\mathbf{L}(O, P, W, R)$, then $(\mathbf{L}, \leq)$ is called a complete concept lattice.

By following the descriptions in the preceding sections, one can notice that the ternary relation R present in these ECG-ontologies will form a many-valued context (or a complete context) as shown in Table 3. By relying on some shareable anchor,these many-valued attributes can be transformed into a binary context using following scaling.

- If some attribute includes (T-inversion, Q-duration) then it's a many-valued attribute and shown by (waveform/T-inversion, waveform/Q-duration),
- If some attribute includes (T-AmplitudeRatio, R-Amplitude) then it's a many-valued attribute and shown by (waveform/T-AmplitudeRatio, waveform/R-Amplitude).

Table 3. A many-valued context extracted from Table 2.

Object	Waveform	Lead	Segment	Reference beat	Complex
AMI	T-inversion Q-duration		ST-depression	QRS	Widening
CAD	T-AmplitudeRatio		ST-depression	QSP	Duration
HYP	P-loss			QRS	Widening
AF	P-loss		RR-inordinance	QRS	Duration
BBB	T-AmplitudeRatio R-Amplitude	(I,V1)		QRS	Duration
WS	T-inversion	(V2,V3)	no-ST-elevation	ST	Duration

As for other attributes, we use the same rules to scale the many-valued attributes into a binary one. The membership values of each binary attribute with their corresponding objects remain the same amount to the many-valued ones.

Conceptualizations of ECG Concepts and Formal Contexts. In practical applications, as we discussed in this text, the conceptualizations underlying different standards are heterogeneous, whose heterogeneity even with respect to the core ECG concepts indicates something else; In reality, we mean that their conceptualizations do not refer directly to the biomedical reality under scrutiny as a shared anchor [21, 22].

To deal with the problems mentioned more adequately, the focus of extracting ECG concepts is mostly on how data and information should be represented in computer and messaging systems, based on top-level domain ontology. Thus, we try to take advantage of ECG-ontologies extracted from AHA/MIT-BIH and SCP-ECG standards and their essential attributes which consist of two parts (the extents and intents). The extents of ECG concepts include the whole objects related to ECG elements, whereas the intents represent the intrinsic properties of the objects.

Table 4 illustrates partial objects of ECG concepts and their essential attributes by FCA, concerned with the representation of what ECG is on the side of the physician. Each row and column of the formal context represents the disease related to ECG diagnosis (extents) and their attributes (intents), according to the clinical effect of different elements on the diagnosis results. If the relation between an object and an attribute exists, then it is denoted as '*'. In any case, we provide a brief description of this ontology, which is sufficient for the purposes of this article. Comparison table between attributes of Table 4 and identifiers is shown as Table 5.

Table 4. Parts of the scaled formal context in binary between entities from the ECG Ontologies of different standards.

Mark	Object	a	b	c	d	e	f	g	h	i	j	k	l	m	n	o
m1	AMI	*	*						*				*		*	
m2	CAD			*					*		*					*
m3	HYP			*									*		*	
m4	AF			*						*			*			*
m5	BBB			*		*	*						*			*
m6	WS	*						*			*		*			

Note: Each letter from "a" to "o" represents attributes in Table 4; * represents criterion satisfied.

The formal concepts of the formal context based on Table 4 can be calculated with a fast algorithm for building lattices [23], in which all concepts are listed as following. The formal concept lattice of partial semantics concepts can be demonstrated in Fig. 2.

C0 = ({m1, m2, m3, m4, m5, m6};);
C1 = ({m2, m4, m5}; {o});
C2 = ({m1, m3, m4, m5}; {l});
C3 = ({m4, m5}; {l, o});

C4 = ({m2, m5}; {c, o});
C5 = ({m3, m4}; {d, l});
C6 = ({m1, m3}; {l, n});
C7 = ({m1, m2}; {h});
C8 = ({m1, m6}; {a});
C9 = ({m5}; {c, e, f, l, o});
C10 = ({m4}; {d, i, l, o});
C11 = ({m3}; {d, l, n});
C12 = ({m2}; {c, h, k, o});
C13 = ({m1; {a, b, h, l, n});
C14 = ({m6}; {a, g, j, m});
C15 = (; {a, b, c, d, e, f, g, h, i, j, k, l, m, n, o}).

Table 5. Partial semantics of concepts between entities in the ECG Ontology and corresponding standards.

Identifier	a	b	c	d	e	f
Attribute	Waveform/ T-inversion	Waveform/ Q-duration	Waveform/ T-AmplitudeRatio	Waveform/ P-loss	Waveform/ R-Amplitude	Lead/ (I, V1)
Identifier	g	h	i	j	k	l
Attribute	Lead/ (V2,V3)	Segment/ ST-depression	Segment/ RR-inordinance	Segment/ no-elevation	Reference beat/QSP	Reference beat/QRS
Identifier	m	n	o			
Attribute	Reference beat/ST	Complex/widening	Complex/duration			

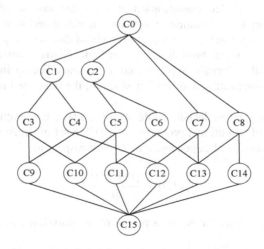

Fig. 2. The general concept lattice for Table 4.

3   Attribute Importance Based on the Inclusion Degree

In order to solve the problem that all the intents of the formal context are viewed as the attributes with equal importance, along with ignoring the relativity of the weighted attributes, the inclusion degree and information entropy, which both characterize differences among the attributes, are introduced by Xiao et al. [24] to design a workflow to build a combined weighted concept lattice in the field of geo-ontologies. The inclusion degree used is to quantify the weights of domain ontology properties, while information entropy mentioned is to differentiate the weights of attributes. It is reasonable to achieve conditional properties, which are significant to get adequate representation of entropy-weighted concept lattice, referenced from the specifications and codes of fundamental geographic information (GB/T 13923-2006). However, how to obtain conditional properties of ECG-ontologies and identify the importance weights of attributes has become a practical and important task in integrating biomedical information.

In this section, based on the analysis in Sect. 2, we consider each of the objects in Table 4 to be a category of pathogenic attributes related to ECG acquisition, for which Table 4 can be viewed as a decision table to calculate the weights of different properties. Besides, to explore the deviation of the importance weights of attributes, we also provide absolute difference from the computed weights of the formal concepts by evaluating the intent's value importance.

3.1   Basic Notes of Rough Set Theory

Rough set theory is a mathematical tool proposed by Professor Pawlak, dealing with incomplete knowledge classification in the fields of data processing [25, 26]. Under the premise of basic knowledge classification, a small amount of prior knowledge is allowed in this theory to text mining, information retrieval and so on. Meanwhile, the vital application of rough set is to obtain attributes of different importance which are determined by vary weights through a decision table. Some necessary knowledge of rough set theory will be briefly implemented as follows. For an in-depth discussion with a rigorous axiomatization and full introduction, the interested reader should refer to [27, 28].

Let $S = (U, A, V, f)$ be an decision table, $A = C \cup D$ is a non-empty set of attributes, in which C is a set of conditional properties and D is a set of decision properties, then $\gamma_C(D)$ denotes the degree of dependence of D with respect to C:

$$\gamma_C(D) = \frac{|POS_C(D)|}{|U|} \tag{4}$$

where $POS_C(D)$ represents the positive region of the partition U/D, $|\cdot|$ represents the number of the set.

For a given decision table $S = (U, A, V, f)$, if $B \subseteq C$, $sig_\gamma(B, C, D)$ denotes the importance of U/D with respect to C:

$$sig_\gamma(B, C, D) = \gamma_C(D) - \gamma_{C-B}(D) \qquad (5)$$

Especially when $B = \{a\}$, the importance of attribute a with respect to decision attribute D is:

$$sig_\gamma(a, C, D) = \gamma_C(D) - \gamma_{C-\{a\}}(D) \qquad (6)$$

3.2 Attribute Importance Based on the Inclusion Degree

Under normal circumstances, one can easily measure the degree of dependence between conditional properties and decision properties using the parameter sig above. However, as we discussed in this text, the parameter may lose the function of other relationship of dependencies, without considering the indirect effect of some single attribute on decision attributes. In order to improve the parameter's ability of describing the importance of property on the properties reduction, XIAO *et al.* [29] put forward the equivalent representation of another definition of the parameter sig, aiming to determine the attribute significance degree based on the attribute dependency degree.

Theorem 1. If $B = \{b_1, b_2, \ldots, b_n\} \subseteq C$, the importance of U/D with respect to $b_i \in B$ is denoted as follows:

$$sig_\gamma'(b_i, B, D) = \frac{1}{n-1} \sum_{1 \le j \le n, j \ne i} \left[sig_\gamma(\{b_j, b_i\}, B, D) - sig_\gamma(b_j, B - \{b_j\}, D) \right] \qquad (7)$$

Proof:

$$sig_\gamma'(b_i, B, D)$$

$$= \gamma_B(D) - \gamma_{B-\{b_i\}}(D)$$

$$= \frac{n-1}{n-1} \left[\gamma_B(D) - \gamma_{B-\{b_i\}}(D) \right] + \sum_{j \ne i} \left[\gamma_{B-\{b_j, b_i\}}(D) - \gamma_{B-\{b_j, b_i\}}(D) \right]$$

$$= \frac{1}{n-1} \sum_{j \ne i} \left[\gamma_B(D) - \gamma_{B-\{b_j, b_i\}}(D) - \gamma_{B-\{b_i\}}(D) - \gamma_{B-\{b_j, b_i\}}(D) \right]$$

$$= \frac{1}{n-1} \sum_{1 \le j \le n, j \ne i} \left[sig_\gamma(\{b_j, b_i\}, B, D) - sig_\gamma(b_j, B - \{b_j\}, D) \right]$$

The importance of a single attribute can be determined by the importance of the attribute set (i.e., the set $\{b_j, b_i\}$ of the attribute to the remaining attributes, respectively) and the importance of the other attributes on the basis of the partial condition attribute set $B - \{b_i\}$. In this process, the major issue is to measure the direct and indirect effects of the condition attributes acted on the decision attributes by using Definition 2. The proper significance degree of attributes is introduced from Ji *et al.* [30].

Definition 2. Let $S = (U, A, V, f)$ be a decision table, in which $A_1, A_2 \subseteq A$, U/A_1 and U/A_2 are partitions of U, defined as $\{X_1, X_2 \ldots X_n\}$ and $\{Y_1, Y_2 \ldots Y_m\}$, respectively; $CON\ (A_1/A_2)$ is denoted as the inclusion degree of U/A_2 to U/A_1:

$$CON(A_1/A_2) = \sum_{0 \le i \le n\ 0 \le j \le m} con(X_i/Y_j) \qquad (8)$$

We assume that the given set, X and Y satisfy a function as follows: when $X \not\subset Y$, con (X/Y) = 0, and if $X \subseteq Y$, con (X/Y) = 1. In the above condition, the range of the function satisfies as following: $0 \le CON\ (A_1/A_2) \le n$. Besides, $CON\ (A_1/A_2)$ can have the maximum value n on condition that A_1 is smaller than A_2.

Definition 3. Let $S = (U, C \cup D, V, f)$ be a decision table, a is a property of conditional properties C; $SIG_\gamma^R(a, C, D)$ denotes the importance of the inclusion degree:

$$SIG_\gamma^R(a, C, D) = \left(|U|^2 sig_\gamma'(a, C, D) + |U|m_a - n + 1 \right) / \left(|U|^2 + |U| + 1 \right) \qquad (9)$$

where

$$m_a = (|POS_a(D)| + CON(a/D)/|U|)/(|U| + 1);$$
$$n_a = (|POS_D(a)| + CON(D/a)/|U|)/(|U| + 1).$$

m_a and n_a represent the influence of the positive region and the inclusion to the property importance, respectively. Obviously, $0 \le m_a, n_a \le 1$.

Theorem 2. The property importance of the inclusion degree $SIG_\gamma^R(a, c, D)$ satisfies the following properties:

1. $0 \le SIG_\gamma^R(a) \le 1$; particularly, $SIG_\gamma^R(a, C, D)$ monotonically increases with m_a and decreases with n_a, So $SIG_\gamma^R(a, C, D)$ could get the minimum 0, when $m_a = 0$ and $n_a = 1$;
2. If $SIG_\gamma^R(a, C, D) = 0$, then $sig_\gamma'(a, C, D) = 0$ must be true, but the opposite is not necessarily the case; If $SIG_\gamma^R(a, C, D) = 1$, then $sig_\gamma'(a, C, D) = 1$ must be true; however, the opposite is not necessarily the case;
3. If $sig_{CD}(a, C, D) > sig_{CD}(b, C, D)$, then $SIG_{CD}(a, C, D) > SIG_{CD}(b, C, D)$, but the opposite is not necessarily the case;
4. If U/a is smaller than U/b, which means $\forall a' \in U/a, \exists b' \in U/b$ makes $X_i \subseteq Y_j$, then $SIG_{CD}(a, C, D) \ge SIG_{CD}\ (b, C, D)$.

4 Construction of an Entropy-Weighted Concept Lattice

Recently, entropy-based formula is widely used for assigning the weight to formal concept of concept lattice. In this section, we are introducing Shannon entropy into concept lattices of ECG-ontologies to reduce the weighted concepts at different fined-grained threshold. To measure the importance of the inclusion degree more precisely, the Shannon entropy and attribute importance are combined to characterize the weights of different properties belonging to the same property in formal context of the decision table.

4.1 Shannon Entropy

Let us consider the given formal context $K_m = (O, P, W, R)$, where $O = \{o_1, o_2,..., o_n\}$ and $P = \{p_1, p_2,..., p_k\}$, then the probability of i-th-object(o_i) processing the corresponding j-th-attribute(p_j) are computed by $P(p_j/o_i)$. The average information weight of attribute p_j can be represented by $E(p_j)$ as the following formula shows:

$$E(p_j) = -\sum\nolimits_{j=1}^{m} P(p_j/o_i) \log_2(P(p_j/o_i)) \tag{10}$$

where m represents the total number of attributes in the given formal context K_m.

4.2 Combined Weights of Properties Based on Attribute Importance and Shannon Entropy

For the analysis of Sect. 2.2, the partial order relation is needed for constructing a concept lattice, especially when we implement semantic integration between ECG data standards. This can be achieved through attributes of reasonable importance. Thus, we proposed another representation of combined weights to quantify the inclusion degree importance of conditional properties more precisely using an entropy-weighted method.

Based on a formal context $K_m = (O, P, W, R)$, the conditional properties of a given decision table $S = (U, C \cup D, V, f)$ can be generated, where $P = \{p_1, p_2,..., p_k\}$ denotes a set of conditional properties. From the aspect of FCA, $SIG(p_j, C, D)$ denotes the inclusion degree importance of a conditional property p_j in the decision table S:

$$SIG(p_j, C, D) = SIG_{\gamma}^{R}(N(p_j, C, D))/|N(p_j, C, D)| \tag{11}$$

where $N(p_j, C, D)$ represents a function to get the value of the conditional property $p_j \in C$ in decision table S. $|N(p_j, C, D)|$ is the number of value of $N(p_j, C, D)$. For instance, considering the formal context of Table 4, we can know that a is the attribute which represents *T-inversion*, and a denotes the attribute of the entities of the concept lattice and $|N(a, C, D)| = 5$.

In order to evaluate the attribute importance of formal concepts, we define a formal context $K_m' = (O, P, W, R, w_c)$ with a set of the combined weights where $P = \{p_1, p_2, ..., p_k\}$ and $w_c(p_j)(1 \leq j \leq k)$ indicates the importance of a attribute p_j in P. The value of $w_c(p_j)$ can be denoted as following:

$$w_c(p_j) = SIG(p_j, C, D) * E(p_j) / \sum\nolimits_{j=1}^{k} SIG(p_j, C, D) * E(p_j) \tag{12}$$

Using Eqs. (1) and (2) in Sect. 2, $w_c(p_j)$ can help to define the weight of the formal concept with the triple $C_w = (A, B, weight (B))$, where $weight (B)$ is the weight of the multi-attribute intent of the formal concept (A, B). Thus, the average weight of the attributes is defined as follows:

$$weight(B) = \sum_1^n (w_c(p_j))/n \tag{13}$$

where $B = b_1 \cup b_2 \cup ... b_n$ and we can assume that $weight (B) = 1$ when $B = \varnothing$.

To explore the deviation of $w_c(p_j)$ from p_j and provide absolute difference from the computed weights of the formal concepts, one needs to evaluate $D(p_j)$ as the intent's value importance, which is denoted as follows:

$$D(p_j) = \sqrt{\sum_{i=1}^n (w_c(p_j) - weight(B))/n} \tag{14}$$

Here, the concept lattice is based on the combined weights of properties produced by attribute importance and Shannon entropy. With the notion $D(p_j)$, we can observe that the entropy-weighted method provides a maximum and minimum deviation of all the formal concepts to analyze the value of their importance. For example, if $n = 1$ then $\sum(w_c(p_j) - weight(B)) = 0$, and similarly one can also compute other values of n.

4.3   Construction of the Reduction Concept Lattice at Different Thresholds

Removal of the Weighted Concept Lattice at Chosen Threshold. For a weighted formal concept $C_w = (A, B, weight (B))$, the semantic granularity threshold θ is defined as the chosen threshold to remove the weighted concept whose weight value is larger than θ, the removal steps of the weighted formal concept are defined in Table 6.

Firstly, the proposed algorithm computes the combined weight of multi-attribute intents through steps 1 to 5. Then, the weight of all formal concepts is computed by adding the sum of their intents using their combined weight through steps 6 to 8. Finally, the weighted formal concept which has the higher weight than the chosen threshold can be removed through steps 9 to 11. Using the algorithm as a basis, we can construct a weighted concept lattice (Fig. 3) transformed from the general concept lattice demonstrated in Fig. 2.

We can observe that the weighted concept lattice shown in Fig. 3 reflects objects m1, m2, m3, m4, m5, m6 as specialization with precise value (also for its covering attributes). This is one of the major advantages of weighted concept lattice to the discovery of domain knowledge, precisely.

Construction of the Weighted Concept Lattice. By using the proposed algorithm to remove the concepts which include more specific characteristics, we concentrate on reducing the weighted concept lattice. With the remaining concepts that contain more common characteristics, the weighted concept lattice can be reconstructed at chosen granularity. What's more, the weighted concept lattice is based on the partial order relation related to the semantic granularity threshold θ, the new lattice is the reduction one which involves the concept with its combined weights value w_c.

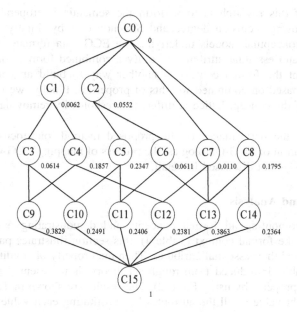

Fig. 3. The weighted concept lattice for Fig. 2.

Table 6. Proposed algorithm for removing weighted formal concepts.

Input: Array [1:s] of all weighted formal concepts
Output: Weighted formal concepts $C_w(s)$ at chosen threshold

---

1.for j=1,...,m where m is the number of multi-attribute intents

2. Compute the probability of any attribute $P(p_j/o_i)$

3. Compute the average information weight of attribute   represented by $E(p_j)$

4. Compute combined weight of multi-attribute intents represented by $w_c(p_j)$

5. **end for**

6. **for** j=1,..., s in which s is the number of weighted formal concepts

7. Compute the weight of attributes in the intent *weight* (C_w)

8. Compute the weight of the weighted formal concept represented by $C_w(s)$

9.   Set the threshold $0 \leq \theta \leq 1$

10.   **if** $(C_w(s) \geq \theta)$

11.   remove the weighted formal concept

12.   **end if**

13.**end for**

---

5   Case-Study

5.1   Case-Study Description

The purpose of this research is to perform the semantic interoperability of ECG-Ontologies by using inclusion degree and Shannon entropy. Firstly, once we have excavated the conceptual models underlying the ECG data formats, we are able to extract objects and essential attributes to unify a combined formal context. We then turn to represent the formal context in another way to build an entropy-weighted concept lattice based on combined weights of properties. Finally, we have performed the reduction of the concept lattice at different thresholds by removing the remaining formal concepts.

To evaluate the performance of the proposed method, our focus is mainly on semantic integration of ECG-Ontologies, the process of mapping ECG-Ontologies into concept lattice was not discussed in detail.

5.2   Results and Analysis

Starting with computing all formal concepts and demonstrating a concept lattice generated from the formal context (Table 4), this section illustrates partial objects of ECG concepts and their essential attributes. Then the property of conditional attributes in a decision table, introduced from rough set theory, is represented to compute the weight of each property by using Eq. (12). The results are shown in Table 7. Table 8 shows the weight value of all the attributes by substituting each value of $SIG_{\gamma}^{R}(property, C, D)$ in Table 7. We took into account the waveform resulting from the process of ECG acquisition as well as the interpretation of this waveform by combining the weights of properties (attribute intents) and the deviation of their importance values based on Eqs. (13) and (14). Table 9 shows the average value of each concept node in Fig. 3.

By using the algorithm of Table 6, the semantic quantity θ can be denoted in different granulations. We can observe that the concepts can be selected with different thresholds in Table 10 according to the range of $Weight(B)$ in Table 9. First, if we set up the semantic quantity $\theta = 0.30$, then the lattice nodes C9, C13 are removed for whose granularity weight values are larger than θ. The reduced weighted concept lattice which is under this θ is shown as Fig. 4. The objects of C9, C13 are "BBB" and "AMI", respectively. This result may confuse some readers for the reasons that "HYP", "AF" and "BBB" should be under the same level of semantic granularity, according to the physician's perspective of ECG diagnosis. Nevertheless, compared to their various attributes, "BBB" and "AMI" include more specific characteristics, such as "waveform/(T-inversion, Q-Duration)" and "waveform/(R-Amplitude, T-AmplitudeRatio)" attributes which only belong to their respective objects. Thus the result is reasonable for those two nodes including more specific characteristics, unsatisfied with the threshold of the intent importance, which indicates that those removed concepts are finer-grained concepts, from the point of granularity computing theory.

Table 7. Weight of each property in decision table.

Inclusion Degree	Waveform	Lead	Segment	Reference	Complex
SIG_γ^R (property, C, D)	0.4425	0.0096	0.0162	0.0816	0.0057

Table 8. Computed weight value of each attribute of the formal context.

Attribute	a	b	c	d	e	f	g	h
P(x)	0.250	0.125	0.250	0.250	0.125	0.125	0.125	0.250
E(x)	0.500	0.375	0.500	0.500	0.375	0.375	0.375	0.500
SIG(x, C, D)	0.0885	0.0885	0.0885	0.0885	0.0885	0.0048	0.0048	0.0054
$w_c(x)$	0.1795	0.1347	0.1795	0.1795	0.1347	0.0073	0.0073	0.0110
Attribute	i	j	k	l	m	n	o	
P(x)	0.125	0.125	0.125	0.500	0.125	0.250	0.375	
E(x)	0.375	0.375	0.375	0.500	0.375	0.500	0.531	
SIG(x, C, D)	0.0054	0.0054	0.0272	0.0272	0.0272	0.0029	0.0029	
$w_c(x)$	0.0082	0.0082	0.0414	0.0552	0.0414	0.0059	0.0062	

Table 9. The intent value of weighted importance and the deviation of concept nodes.

Node	Intent	Average value	Weight(B)	D(B)
C0	∅	1	1	0
C1	o	0.0062	0.0062	0
C2	l	0.0552	0.0552	0
C3	lo	0.0614	0.0614	0.0314
C4	co	0.1857	0.1857	0.0962
C5	dl	0.2347	0.2347	0.1226
C6	ln	0.0611	0.0611	0.0313
C7	h	0.0110	0.0110	0
C8	a	0.1795	0.1795	0
C9	ceflo	0.3829	0.3829	0.0876
C10	dilo	0.2491	0.2491	0.1255
C11	dln	0.2406	0.2406	0.1196
C12	chko	0.2381	0.2381	0.1230
C13	abhln	0.3863	0.3863	0.0912
C14	agjm	0.2364	0.2364	0.1238

Similarly, if the semantic quantity is set to $\theta = 0.10$, then the lattice nodes C4, C5, C8, C9, C10, C11, C12, C13, C14 are removed, for whose semantic granularity is larger than 0.10. The reduced weighted concept lattice is plotted in Fig. 5. It is obvious to see that the whole concept nodes which contain a single object are removed at this level of semantic granularity. The result is no surprise as any formal concept which contains more than one object often remains common attributes and discards non-common attributes.

Table 10. The reduced weighted concept lattice in Table 9 at different weights.

Weight(B)	θ	Obtained formal concepts
1	$0.3863 < \theta \leq 1$	∅
0.3863	$0.3829 < \theta \leq 0.3863$	C13
0.3829	$0.2491 < \theta \leq 0.3829$	C13, C9
0.2491	$0.2406 < \theta \leq 0.2491$	C13, C9, C10
0.2406	$0.2381 < \theta \leq 0.2406$	C13, C9, C10, C11
0.2381	$0.2364 < \theta \leq 0.2381$	C13, C9, C10, C11, C12
0.2364	$0.2347 < \theta \leq 0.2364$	C13, C9, C10, C11, C12, C14
0.2347	$0.1857 < \theta \leq 0.2347$	C13, C9, C10, C11, C12, C14, C5
0.1857	$0.1795 < \theta \leq 0.1857$	C13, C9, C10, C11, C12, C14, C5.C4
0.1795	$0.0614 < \theta \leq 0.1795$	C13, C9, C10, C11, C12, C14, C5.C4, C8
0.0614	$0.0611 < \theta \leq 0.0614$	C13, C9, C10, C11, C12, C14, C5.C4, C8, C3
0.0611	$0.0552 < \theta \leq 0.0611$	C13, C9, C10, C11, C12, C14, C5.C4, C8, C3, C6
0.0552	$0.0110 < \theta \leq 0.0552$	C13, C9, C10, C11, C12, C14, C5.C4, C8, C3, C6, C2
0.0110	$0.0062 < \theta \leq 0.0110$	C13, C9, C10, C11, C12, C14, C5.C4, C8, C3, C6, C2, C7
0.0062	$0 < \theta \leq 0.0062$	C13, C9, C10, C11, C12, C14, C5.C4, C8, C3, C6, C2, C7, C1

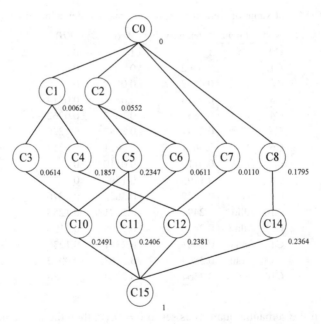

Fig. 4. The reduced weighted concept lattice (θ = 0.30).

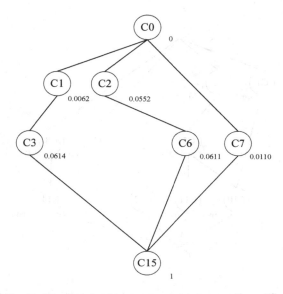

Fig. 5. The reduced weighted concept lattice ($\theta = 0.10$).

When comparing the results between Figs. 4 and 5, we can see that the number of concept nodes under $\theta = 0.10$ is fewer than that under $\theta = 0.30$, whereas the set of the reduced concept lattice under $\theta = 0.10$ is a subset of that under $\theta = 0.30$. Based on the analysis above, one can draw a conclusion that the Hasse Diagram of the reduced weighted lattice is getting gradually simplified with the decrease of the semantic quantity threshold (θ). Meanwhile, the process of reducing the combined weighted lattice is a stepwise refinement process in light of various levels of granularity.

Finally, the quantity η ($0 \leq \eta \leq 1$) is defined as the threshold of the deviation of the intent importance. By using the terms of D(B) in Table 9, if $\eta = 0.10$ was set up as the deviation threshold when ($\theta = 0.30$), then C3, C4, C6, C9, and C13 are moved for whose value of D(B) was lower than η, that's to say, these nodes could be deleted according to the intent importance thresholds specified by domain experts, even if the intent weight values of these nodes (C4, C9, and C13) are greater than the predefined threshold (θ). The reduced weighted concept lattice when ($\theta = 0.30$ and $\eta = 0.10$) is shown in Fig. 6, which is simplified compared to Fig. 6. Besides, by the methods of the deviance analysis, if the greater is the absolute difference between the weight values of attributes intent, the greater is the deviation value of D(B), and the contrary has also been set up. For instance, the deviation value of C4 is greater than that of C3 in Table 9, and then we can draw a conclusion that the weight value difference between "c" and "o" is greater than that between "l" and "o". Based on the analysis above, we should first retain some single nodes whose weight values of attributes intent have greater differences. Afterwards, other nodes with smaller difference are removed by improving the deviance value in the course of reducing the weighted concept lattice. Thus, it's vital to select an appropriate threshold especially when we reduce the complexity of the algorithm scale and promote the level of knowledge expression.

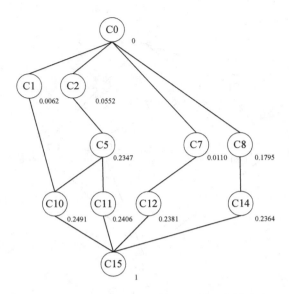

Fig. 6. The reduced weighted concept lattice ($\theta = 0.30$ and $\eta = 0.10$).

6   Conclusions and Outlook

In this paper, we aimed at constructing a combined weighted concept lattice of ECG-Ontologies. In this process, the major issue is to reduce the number of formal concepts and the size of weighted concept lattice. To overcome it, we first matched a correspondent entity of the ECG-Ontologies by excavating the conceptual models underlying the ECG data formats, and then expressed the ontology and its essential attributes as the extents and the intents of ECG concepts, which are formalized to be a unified formal context. Secondly, we introduced a method for computing the weight of given formal concepts by combing the inclusion degree and Shannon entropy, the purpose of which is to characterize differences of various attributes in this unified formal context. Thirdly, we built a weighted concept lattice based on the unified formal context and described the combined weight of each concept node in detail to reduce the less valued and redundant concepts. Finally, we reduced the intent of the weighted concept lattice by gradually lowering the value of granularity threshold and removing concepts with a larger granularity than the given threshold. The reduced lattice could provide adequate description of ECG formal concepts with regular hierarchical order visualization. The results of experiments showed that the proposed method proved to be feasible and effective in the aspect of integrating heterogeneous ECG-Ontologies.

In this study we have made the decision table based on the conceptual model resulting from excavation of AHA/MIT-BIH and SCP-ECG standards, represented as extents and their attributes (intents), among which the decision property can be reasonable as prior classification knowledge. Consequently, in the future, we will focus on enlarging the number of decision properties to merge multiple properties of different ECG domains. Besides, as discussed above, a suitable threshold relying on domain experts may become a bottleneck in controlling the simplicity, so our future goal also includes setting up threshold via self-adaptive behavior.

Acknowledgments. The author would like to thank all the anonymous reviewers and the Editor-in-Chief for their valuable time to improve our paper. This work was partly supported by the Key Research Funds of Anhui Provincial Department of Education (Nos. SK2018A1064, SK2018A1072 and KJ2018A1007), the Teaching Research Major Projects of Anhui Province (2018jyxm1446), and the Demonstration experiment training center of Anhui Province (2018sxzx58).

References

1. Goldberger, A.L., Amaral, L.A., Glass, L.: PhysioBank, PhysioToolkit, and PhysioNet: components of a new research resource for complex physiologic signals. Circulation **101**(23), E215 (2000)
2. CEN/TC-251. SCP Document CEN/TC-251 N02-15 (2002). http://www.centc251.org/. Accessed August 2006
3. Stockbridge, N., Brown, B.D.: Annotated ECG waveform data at FDA1. J. Electrocardiol. **37**, 63–64 (2004)
4. Smith, B., Ashburner, M., Rosse, C.: The OBO Foundry: coordinated evolution of ontologies to support biomedical data integration. Nat. Biotechnol. **25**(11), 1251–1255 (2007)
5. Guarino, N.: Formal ontology and information systems. In: Guarino, N. (ed.) Proceedings of 1st International Conference on Formal Ontology and Information Systems, Trento, Italy, pp. 3–15 (1998)
6. Smith, B., Welty, C.: Ontology: towards a new synthesis. In: Smith, B., Welty, C. (eds.) Proceedings of 2nd International Conference on Formal Ontology in Information Systems (FOIS), pp. 3–9. ACM Press, New York (2001)
7. Abdelkader, M., Mimoun, M., Djeloul, B.: Locating services in legacy software: information retrieval techniques, ontology and FCA based approach. Wseas Trans. Comput. **11**(1), 19–26 (2012)
8. Fu, G.: FCA based ontology development for data integration. Inf. Process. Manag. **52**(5), 765–782 (2016)
9. Fang, K., Chang, C., Chi, Y.: Using formal concept analysis to leverage ontology-based Acu-Point knowledge system. In: Zhang, D. (ed.) ICMB 2008. LNCS, vol. 4901, pp. 115–121. Springer, Heidelberg (2007). https://doi.org/10.1007/978-3-540-77413-6_15
10. Jiayun, Y., Yang, J., Feng, M.: FCA-based research on hierarchy relationship of medical terms. Inf. Res. (2017)
11. Kim, I.-C.: FCA-based ontology augmentation in a medical domain. In: Karagiannis, D., Reimer, U. (eds.) PAKM 2004. LNCS (LNAI), vol. 3336, pp. 408–413. Springer, Heidelberg (2004). https://doi.org/10.1007/978-3-540-30545-3_38
12. Cimino, J.J.: Desiderata for controlled medical vocabularies in the twenty-first century. Methods Inf. Med. **37**(4–5), 394–403 (1998)
13. Zhang, S.: A completeness analysis of frequent weighted concept lattices and their algebraic properties, Data Knowl. Eng. **81–82**(4), 104–117 (2012)
14. Moody, G.B.: WFDB programmer's guide. Version 10.4.19 (2009). http://www.physionet.org/physiotools/wpg/. Accessed March 2009

15. Stockbridge, N., Brown, B.: Annotated ECG waveform data at FDA 1. J. Electrocardiol. **37**, 63–64 (2004)
16. Medvidovic, N., Rosenblum, D.S., Redmiles, D.F.: Modeling software architectures in unified modeling language. ACM Trans. Softw. Eng. Methodol. **11**(1), 2–57 (2002)
17. Martins, J.L., Fox, K.F., Wood, D.A.: Rapid access arrhythmia clinic for the diagnosis and management of new arrhythmias presenting in the community: a prospective, descriptive study. Heart **90**(8), 877–881 (2004)
18. Wille, R.: Concept lattices and conceptual knowledge systems. Comput. Math Appl. **23**, 493–515 (1992)
19. Kumar, C.A.: Knowledge discovery in data using formal concept analysis and random projections. Int. J. Appl. Math. Comp. **21**, 745–756 (2011)
20. Kwon, O., Kim, J.: Concept lattices for visualizing and generating user profiles for context-aware service recommendations. Expert Syst. Appl. **36**, 1893–1902 (2009)
21. Johansson, I.: Bioinformatics and biological reality. J. Biomed. Inf. **39**(3), 274–287 (2006)
22. Smith, B.: From concepts to clinical reality: an essay on the benchmarking of biomedical terminologies. J. Biomed. Inf. **39**(3), 288–298 (2006)
23. Yang, K.M., Kim, E.H., Hwang, S.H.: Fuzzy concept mining based on formal concept analysis. Int. J. Comput. **2**(3), 279–290 (2008)
24. Xiao, J., He, Z.: A concept lattice for semantic integration of geo-ontologies based on weight of inclusion degree importance and information entropy. Entropy **18**(399), 1–16 (2016)
25. Ohrn, A., Rowland, T.: Rough sets: a knowledge discovery technique for multifactorial medical outcomes. Am. J. Phys. Med. Rehabil. **79**(1), 100 (2000)
26. Yao, Y.Y.: Three-way decisions with probabilistic rough sets. Inf. Sci. **180**, 341–353 (2010)
27. Pawlak, Z.: Rough Sets. Int. J. Comput. Inf. Sci. **11**, 341–356 (1982)
28. Zhang, W.X., Wu, W.Z., Liang, J.Y.: Theory and Method of Rough Sets. Science Press, Beijing (2001). (in Chinese)
29. Jinsen, X., Limin, S.: Improved attribute significance degree based on rough set. Comput. Eng. Appl. **53**(3), 174–176 (2017). (in Chinese)
30. Ji, J., Wu, G.X., Li, W.: Significance of attribute based on inclusion degree, J. Jiangxi Norm. Univ. (Nat. Sci.) **33**, 656–660 (2009). (in Chinese)

Minimizing the Spread of Rumor Within Budget Constraint in Online Network

Songsong Mo, Shan Tian, Liwei Wang$^{(\boxtimes)}$, and Zhiyong Peng

School of Computer Science, Wuhan University, Wuhan, Hubei, China
{songsong945,tianshan14,liwei.wang,peng}@whu.edu.cn

Abstract. The spread of rumor in online network results in undesirable social effects and even leads to economic losses. To overcome this problem, a lot of work studies the problem of rumor control which aims at limiting the spread of rumor. Unfortunately, all previous work only assumes that users are passive receivers of rumors even if the users can browse the rumors on their own, or the cost of the 'anti-rumor' node is uniform although it is impossible that the price of broadcasting information on Baidu's homepage is the same as personal homepage's. Considering the above problems, in this paper, we study the Rumor Control within Budget Constraint (RCBC) problem. Given a node-weighted graph $G(V, E)$ and a budget B, it aims to find a protector set P, which can minimize the spread of rumor set R in the online network, and the total cost of node in P does not exceed budget B. In consideration of the rumor spread via users' browsing behaviors, we model the rumor propagation based on the random walk model. To solve this problem efficiently, we propose two greedy algorithms that can approximate RCBC within a ratio of $\frac{1}{2}(1 - 1/e)$. To improve the efficiency of them further, we devise a PreSample method to eliminate nodes that can't access R by T-length random walk. Experiments on real datasets have been conducted to verify the efficiency, effectiveness, memory consumption and scalability of our methods.

Keywords: Social network · Rumor control · Random walk · Budget constraint

1 Introduction

With the increasing popularity of online networks, World Wide Web, social networks and peer-to-peer networks have become the most commonly utilized vehicles for information propagation and changed people's lifestyle greatly. However, the ease of information propagation is a double-edged sword. It also can quickly spread rumors and misinformation, which results in undesirable social effects and even leads to economic losses [4,14,16,17]. For instance, the fake tweet about the earthquake in Ghazni province in August 2012 made thousands of people leave their home in panic and be afraid of returning back home [7]. And the rumor

© Springer Nature Singapore Pte Ltd. 2019
X. Sun et al. (Eds.): NCTCS 2019, CCIS 1069, pp. 131–149, 2019.
https://doi.org/10.1007/978-981-15-0105-0_9

of explosion in White House in 2013 caused \$130 billion loss on the stock market[1]. Therefore, minimizing the spread of rumor in online network is a crucial problem.

Against this backdrop, a lot of work studies the problem of rumor control which aims to minimize the spread of rumor on social network [1–3,5,8,13]. Unfortunately, most previous work only assumes that users are passive receivers of rumors even if the users can browse the rumors on their own. Therefore, in this study, we assume that users will actively encounter/contact the rumors via their browsing behaviors, i.e., keyword search, social browsing, etc, which can be modeled by random walk model [19]. However, in [19], they assume the cost of each protector is uniform, which is inconsistent with the non-uniform cost of deploying information on different web pages in real life. For example, a company wants to deploy some 'anti-rumor' information to minimize the spread of the fake news about their company on the World Wide Web. It is impossible that the price of broadcasting information on Baidu's homepage is the same as personal homepage's.

Motivated by this, we study the problem of minimizing the spread of rumor within budget constraint, and call it Rumor Control within Budget Constraint (RCBC) problem. Suppose that an online network is represented by a node-weighted graph $G(V, E)$. Given a rumor set $R \in V$ and a budget B, RCBC aims to find a protector set $P \in V \backslash R$ to minimize the spread of the rumor set R as much as possible such that the total cost of node in P does not exceed budget B. In order to solve this problem, we first analyze the monotonicity and submodularity of the objective function of RCBC (Sect. 2.3). Due to the fact that the problem of maximizing a submodular function is generally NP-hard [11], we resort to developing approximate algorithms to solve it efficiently.

To this end, we first devise a Monte Carlo (MC) based greedy (GreedySel), which is extended from the algorithm for the general Budgeted Maximum Coverage (BMC) problem [10] and can provide $\frac{1}{2}(1 - 1/e)$ approximation factor for RCBC, as baseline. To further speed up GreedySel, we propose an efficient index structure and give an index-based method (IndexSel) to accelerate each round of unit marginal gain calculation. Moreover, we devise an pre-sample method to eliminate nodes that can't access R by T-length random walk, which can speed up GreedySel and IndexSel further.

In summary, we make the following contributions.

- We propose and study the RCBC problem, and analyze the monotonicity and submodularity of the objective function of RCBC.
- To solve the RCBC problem, we present a Monte Carlo based greedy algorithm (GreedySel) as the baseline solution. Moreover, we diverse a index based greedy method (IndexSel), which can speed up GreedySel greatly.
- We propose a PreSample method to further eliminate nodes that can't access R by T-length random walk, which can accelerate GreedySel and IndexSel further.

---

[1] https://www.dailymail.co.uk/sciencetech/article-3090221.

- We conduct extensive experiments on three real-world datasets. The results validate the effectiveness, efficiency, memory consumption and scalability of our methods.

2 Preliminary

In this section, we first formally define the influence model and then give the definition of RCBC. In the following, we analyze the monotonicity and submodularity of the objective function of RCBC. In the end, we discuss the most relevant literature to this paper. Important notations used in our paper are presented in Table 1.

Table 1. Notations for problem formulation and solutions

Symbol	Description
$G(V, E)$	A node-weighted graph
w_u	An instance of a random walk starting from u
T	A given threshold to bound the length of w_u
B	A given budget
R, P	$R \subset V$ is a rumor set, $P \subset V$ is a protector set, and $R \cap P = \phi$
$I_{w_u}(P|R)$	The value of probability that P blocks the influence of R to u for w_u
$I_u(P|R)$	$I_u(P|R) = E[I_{w_u}(P|R)]$ for any w_u
$\mathcal{G}(P|R)$	The object function of our problem (the block degree of P)
$u.\mathcal{D}$	The block degree of node u

2.1 A Random Walk Based Influence Block Model

We assume that a social network is a node-weighted graph $G = (V, E)$ with $n = |V|$ nodes and $m = |E|$ edges. Given a node $u \in V$, a browsing process starting from u can be represented by a random walk w_u [15]. In particular, w_u picks a neighbor of u by the probability of p_{uv} (where $p_{uv} = 1/d_u$ and d_u is the out-degree of u) and move to this neighbor, and then follows this way recursively.

Here we say that u hits (or is influenced by) set S at the time step t, if w_u first visits set S by a t-hop jump. It is worth noting that t should not be very large in real world, as most of social media users only browse a small number of pages in each day. Therefore, we can use a threshold T to bound the hitting time t for any nodes and sets. Based on t, we introduce the concept of influence block as follows.

Definition 1 (Influence Block). *Let P and R be two disjoint sets in V. For a random walk w_u, we define that P block the influence of R to u, if w_u visits P before R, w.r.t. w_u first visits P and R at the time step t_1 and t_2 respectively, and $t_1 < t_2$.*

2.2 Problem Definition

Given such an influence block model, let $I_u(P|R) = E[I_{w_u}(P|R)]$ denote the expected value of possibility that P blocks the influence of R to u. Here $I_{w_u}(P|R)$ denote the possibility that P blocks R's influence to u in w_u. Based on $I_u(P|R)$, the problem of Rumor Control within Budget Constraint (RCBC) can be described as the follows.

Definition 2 (Problem Definition). *Given a weighted graph $G = (V, E, c)$, an initial set $R \subset V$ and a budget B, RCBC is dedicated to finding a set $P \subset V \backslash R$, which can maximize $\mathcal{G}(P|R) = \sum_{u \in V \backslash R} I_u(P|R)$ such that $c(P) = \sum_{v \in P} v.c < B$.*

Here $\mathcal{G}(P|R)$ denotes the block degree of P relative to R. $c(P)$ and $v.c$ denote the total cost of protector set P and the cost of node v, respectively. It is worth noting that our problem does not consider the influence cascade for simplicity, which means that the initial node will not affect the neighbor node actively.

2.3 The Monotonicity and Submodularity of the RCBCProblem

Theorem 1. *The objective function $\mathcal{G}(P|R)$ of RCBC is monotone and submodular.*

Proof. We do not prove the monotonicity of $\mathcal{G}(P|R)$ since it is very straightforward. It remains to show that $\mathcal{G}(P|R)$ is submodular.

Let $S \subseteq T \subset V$ and v (here $v.c < B - c(T)$) be a node in $V \backslash T$. According to [12], $\mathcal{G}(S|R)$ is submodular if it satisfies:

$$\mathcal{G}(S \cup v|R) - \mathcal{G}(S|R) \geq \mathcal{G}(T \cup v|R) - \mathcal{G}(T|R).$$

To facilitate the proof, we define $S_v = S \cup v$ and $\mathcal{G}_v(S) = \mathcal{G}(S \cup v|R) - \mathcal{G}(S|R)$. Let $\mathcal{W}_u$ denote the set of all possible random walk from u. Intuitively, $I_u(P|R)$ can be rewritten as $\sum_{w_u \in \mathcal{W}_u} I_{w_u}(P|R)\mathcal{P}(w_u)$, where $\mathcal{P}(w_u)$ denotes the probability of w_u. Then, we have:

$$\begin{aligned}
\mathcal{G}_v(S) &= \mathcal{G}(S_v|R) - \mathcal{G}(S|R) \\
&= \sum_{u \in V \backslash R} I_u(S_v|R) - \sum_{u \in V \backslash R} I_u(S|R) \\
&= \sum_{u \in V \backslash R} \sum_{w_u \in \mathcal{W}_u} I_{w_u}(S_v|R)\mathcal{P}(w_u) - \sum_{u \in V \backslash R} \sum_{w_u \in \mathcal{W}_u} I_{w_u}(S|R)\mathcal{P}(w_u) \\
&= \sum_{u \in V \backslash R} \sum_{w_u \in \mathcal{W}_u} (I_{w_u}(S_v|R) - I_{w_u}(S|R))\mathcal{P}(w_u).
\end{aligned}$$

Hence, we have:

$$\begin{aligned}
\mathcal{G}_v(S) - \mathcal{G}_v(T) &= \sum_{u \in V \backslash R} \sum_{w_u \in \mathcal{W}_u} (I_{w_u}(S_v|R) - I_{w_u}(S|R))\mathcal{P}(w_u) \\
&\quad - \sum_{u \in V \backslash R} \sum_{w_u \in \mathcal{W}_u} (I_{w_u}(T_v|R) - I_{w_u}(T|R))\mathcal{P}(w_u) \\
&= \sum_{u \in V \backslash R} \sum_{w_u \in \mathcal{W}_u} \mathcal{P}(w_u) \\
&\quad * ((I_{w_u}(S_v|R) - I_{w_u}(S|R)) - (I_{w_u}(T_v|R) - I_{w_u}(T|R)))
\end{aligned} \tag{1}$$

To show the submodularity of RCBC, we first prove Eq. 2.

$$((I_{w_u}(S_v|R) - I_{w_u}(S|R)) - (I_{w_u}(T_v|R) - I_{w_u}(T|R))) \geq 0 \qquad (2)$$

Since $S \subseteq T$, we have $I_{w_u}(S|R) \leq I_{w_u}(T|R)$. On one hand, if v can block the influence of R to u in w_u, we have $I_{w_u}(S_v|R) = I_{w_u}(T_v|R)$. Thus, $((I_{w_u}(S_v|R) - I_{w_u}(S|R)) - (I_{w_u}(T_v|R) - I_{w_u}(T|R))) \geq 0$. On the other hand, if v can not block the influence of R to u in w_u, we have $I_{w_u}(S_v|R) = I_{w_u}(S|R)$ and $I_{w_u}(T_v|R) = I_{w_u}(T|R)$. Hence, $((I_{w_u}(S_v|R) - I_{w_u}(S|R)) - (I_{w_u}(T_v|R) - I_{w_u}(T|R))) = 0$. These discussions show the correctness of Eq. 2.

Based on Eqs. 1 and 2, we have $\mathcal{G}_v(S) - \mathcal{G}_v(T) \geq 0$ and thus $\mathcal{G}(P|R)$ is a submodular function.

2.4 Related Work

In the following, we discuss the most relevant literature to our problem.

One closely related work is the proactive rumor control in online networks [19], which only studies one specific case of our problem, assuming that the cost of all nodes in G is uniform. But in fact, we cannot ignore the cost information of the nodes. For example, it is impossible that the price of broadcasting information on Baidu's homepage is the same as personal homepage's. Therefore, we study the rumor control within budget constraint problem.

Competitive influence maximization (CIM) and influence block (IBK) are two other problem that are close to our problem. CIM aims to identify a set of target seed nodes (or protectors) who will spread an 'anti-rumor' to limit the scale of rumor propagation [2,3,5]. Carnes et al. [5] and Bharathi et al. [2] study competitive influence diffusion under the extension of the IC model and show that the problem of maximizing the influence of one campaign is NP-hard and submodular, while Borodin et al. [3] studies the similar problem under the LT model. Different with CIM, IBK tries to limit the influence of rumor by blocking some nodes or links in a network [1,8,13]. Their strategies of the seed selection are mainly based on their connectivity, such as degree [1,13], pagerank [8], and betweenness [8]. Our problem is essentially different from the above work for the following reason. Both CIM and IBK assume that the information (or rumors) propagations are driven by the effect of word-of-mouth, and they use Independent Cascade model (IC) and Linear Threshold model (LT) to simulate the spread of rumors. However, our problem assumes that rumors spread via browsing behaviors and uses a random walk model to describe the influence spread of rumors.

3 Approximate Solutions

In general, the problem of submodular function maximization is NP-hard [11]. Therefore, we turn to find approximate solutions in polynomial time. In this section, we first present a Monte Carlo based greedy method (GreedySel), that is extended from the algorithm for the general Budgeted Maximum Coverage (BMC) problem [10], as baseline. For GreedySel, the main bottleneck is to repeatedly scan the sample set for marginal gain calculation. To overcome this problem, we propose an efficient index structure and give an index-based method (IndexSel) to accelerate each round of unit marginal gain calculation.

3.1 A Basic Greedy Method

The core idea of GreedySel is to select the node u which maximizes the unit marginal gain, i.e., $(\mathcal{G}(P \cup \{u\}|R) - \mathcal{G}(P|R))/u.c$, to a candidate solution set P, until the budget is exhausted. The pseudo code of GreedySel is presented in Algorithm 1. It first initializes P as empty set and $V \leftarrow V\backslash R$. Next, it finds a set P according to the greedy heuristic (Lines 1.6 to 1.10). The set P is the first candidate of final output. The second candidate is a single set H for which block degree is maximized and cost does not exceed the budget. In the end, it outputs the candidate solution having the maximum block degree.

According to [10], Algorithm 1 achieves $(1 - 1/\sqrt{e})$ approximation factor for RCBC. But in [18], the approximate ratio of Algorithm 1 is corrected to $\frac{1}{2}(1 - 1/e)$.

Complexity Analysis. Let L be the number of random walk simulations for each node. Suppose that the length of random walk is bounded by T, thus the sampling set is generated in $O(nLT)$ time. In each iteration, Algorithm 1 needs to scan the sampling set once to pick the node with the highest unit marginal gain. Therefore, the time and space complexity of Algorithm 1 are $O(n^2 LT)$ and $O(nLT)$ respectively.

3.2 Index Based Greedy Method

As we analyzed above, the most expensive part of Algorithm 1 is to compute $\mathcal{G}(P|R)$, which needs to scan the sampling set repeatedly. To further accelerate GreedySel, we devise an effective bidirectional mapping index based on the sampling set to avoid scanning the full set. At the same time, we use heap to accelerate the process of selecting the node with the largest unit marginal gain.

Bidirectional Mapping Index $\mathcal{B}$. The index is shown in Fig. 1. In this figure, we link a node n_i in Node List and a random walk w_j in Random Walk List by a bidirectional pointer, if the node can block the influence of R to n_i in w_j. The random walk set RS of node v contains random walk w and the node set NS of w_j contains n_i if there is a bidirectional pointer between w_j and n_i. If any node $n_i \in w_j.NS$ and $n_i \in P$, we set $w_j.isBlocked = True$, otherwise $w_j.isBlocked = False$. The block degree of node n_i ($n_i.\mathcal{D}$) denotes the number of random walk in $n_i.RS$ that are not blocked.

Algorithm 1. GreedySel(G, R, B)

1.1 **Input:** a graph G, a rumor set R and a budget B
1.2 **Output:** a protector set P such that $c(P) < B$
1.3 Run L T-random walks for each node in $V \backslash R$
1.4 $Is(R) \leftarrow$ all the random walks influenced by R
1.5 Initialize P as an empty set and $V \leftarrow V \backslash R$.
1.6 **repeat**
      /* $\mathcal{G}(P|R)$ is computed based on Monte Carlo simulations $Is(R)$.   */
1.7   Select $u \leftarrow \arg\max_{v \in V}((\mathcal{G}(P \cup \{v\}|R) - \mathcal{G}(P|R))/v.c)$
1.8   **if** $c(P) + u.c \leq B$ **then**
1.9 | $P \leftarrow P \cup \{u\}$
1.10 $V \leftarrow V \backslash \{u\}$
1.11 **until** $V = \phi$
1.12 $H \leftarrow \arg\max_{v \in V}(\mathcal{G}(v|R))$ and $v.c \leq B$
1.13 **if** $\mathcal{G}(H|R) \leq \mathcal{G}(P|R)$ **then**
1.14 | **return** P
1.15 **else**
1.16 | **return** H

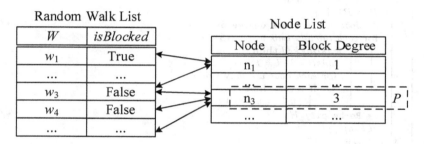

Fig. 1. Bidirectional mapping index $\mathcal{B}$.

Based on this bidirectional mapping index, the pseudo code of the index based method IndexSel is presented in Algorithm 2. It first builds $\mathcal{B}$ based on sampling set (Lines 2.3 to 2.5). After that, it computes the block degree of each node and initializes a max-heap $\mathcal{H}$ according to the unit block degree of each node (Lines 2.7 to 2.9). Based on $\mathcal{H}$, it picks the node v with highest unit marginal gain into P iteratively if $c(P) + v.c \leq B$. After each round of selection, it updates the block degree of the rest node and re-heapifies the heap $\mathcal{H}$ (Lines 2.10 to 2.22). It is worth noting that this algorithm updates the block degree of the rest nodes by searching $\mathcal{B}$ to avoid computing all the remaining nodes (Lines 2.14 to 2.19). In the end, it outputs the candidate solution having the maximum block degree.

Complexity Analysis. In the initialization phase of $\mathcal{H}$, Algorithm 2 takes $O(nLT)$ time to compute the block degree of nodes by scanning the sampling set and $O(n)$ time to build the heap $\mathcal{H}$. In each round of node selection, it needs to update the block degree of nodes and re-heapify $\mathcal{H}$. According to flag $isBlocked$, it ensures that the updating operations totally takes $O(nLT)$ time. Moreover, heapifying a heap once takes $O(\log n)$ [6]. Therefore, the time complexity of Algorithm 2 is $O(nLT \log n)$.

Algorithm 2. IndexSel(G, R, B)

2.1 **Input:** a graph G, a rumor set R and a budget B
2.2 **Output:** a protector set P such that $c(P) < B$
     /* Sampling and Indexing   */
2.3 Run L T-random walks for each node in $V \backslash R$
2.4 $Is(R) \leftarrow$ all the random walks influenced by R
2.5 Precompute $\mathcal{B}$ to index $Is(R)$
2.6 Initialize $P \leftarrow \phi$ and $V \leftarrow V \backslash R$
     /* Get the block number for each candidate.   */
2.7 **for** *each node $v \in V$* **do**
2.8 | Compute $v.\mathcal{D}$ by scan $Is(R)$
2.9 Build max-heap $\mathcal{H}$ for $v \in V$ with the unit block degree $(v.\mathcal{D}/v.c)$
2.10 **repeat**
2.11 | $v \leftarrow \mathcal{H}.pop()$
2.12 | **if** $c(P) + v.c \leq B$ **then**
2.13 | | $P \leftarrow P \cup \{v\}$
     | /* Update the block degree of node.   */
2.14 | | **for** *each random walk $w \in v.RS$* **do**
2.15 | | | **if** $!w.isBlocked$ **then**
2.16 | | | | $w.isBlocked = True$
2.17 | | | | **for** *each node $u \in w.NS$* **do**
2.18 | | | | | **if** $u \notin P$ **then**
2.19 | | | | | | $u.\mathcal{D} = u.\mathcal{D} - 1$
     | /* Re-heapify $\mathcal{H}$.   */
2.20 | | $\mathcal{H}.heapify()$
2.21 | $V \leftarrow V \backslash \{v\}$
2.22 **until** $V = \phi$
2.23 $H \leftarrow \arg\max_{v \in V}(\mathcal{G}(v|R))$ and $v.c \leq B$
2.24 **if** $\mathcal{G}(H|R) \leq \mathcal{G}(P|R)$ **then**
2.25 | **return** P
2.26 **else**
2.27 | **return** H

---

4 Pre-sampling Based Solutions

As we show in Sect. 3, we sample every point in G and calculate the marginal gain by scanning the sample set, which leads to more memory consumption and the time overhead of scanning the sample set. However, some nodes can't access the rumor set R, due to the limitations of network community and random walk length T. Motivated by this, we devise an pre-sample method to eliminate nodes that can't access R by T-length random walk, which can speed up GreedySel and IndexSel further.

4.1   Pre-sampling Stage

In this section, we introduce a method to eliminate nodes that can't access R by T-length random walk by sampling. Our analysis uses the Hoeffding's inequality [9]:

Theorem 1 (Hoeffding's Inequality). *Let* $Z_1, \ldots, Z_m$ *be independent bounded random variables with* $Z_i \in [a, b]$ *for all* i *with a mean* μ., *where* $-\infty < a \leq b < \infty$. *Then*

$$\Pr\left(\frac{1}{m}\sum_{i=1}^{m}(|Z_i - \mu|) \geq t\right) \leq \exp\left(-\frac{2mt^2}{(b-a)^2}\right)$$

for all $t \geq 0$.

Let P_u denote the possibility of u access R by a T-length random walk. To compute P_u, we independently run r T-length random walks starting from node $u \notin P$, and take the average hits $(\frac{1}{r}\sum_{i=1}^{r}(X_i)$, here if random walk i hit R, $X_i = 1$, otherwise, $X_i = 0$) as the estimator. The proposed sampling process is equivalent to a simple random sampling with replacement, thus $\frac{1}{r}\sum_{i=1}^{r}(X_i)$ is an unbiased estimation of P_u. By applying this to Theorem 1, we have

$$\Pr\left(|\frac{1}{r}\sum_{i=1}^{r}X_i - \mu| \geq t\right) \leq \exp\left(-2rt^2\right)$$

for all $t \geq 0$. Hence, we can return a $(1-t)$-approximate solution for P_u with at least $(1 - \exp(-2rt^2))$ probability. In this paper, we set $t = 0.05$ and $r = 500$.

The pseudo code of pre-sampling stage is presented in Algorithm 3. It first computes P_u for each node $u \in V \backslash R$ by running r T-random walks for each node in $V \backslash R$. After computing P_u for each node $u \notin R$, we remove node v and the edges associated with it if $P_v < 0.1$. In the end, it outputs the processed graph.

4.2   Putting It Together

In summary, as presented in Algorithms 4 and 5, our pre-sampling based solutions work as follows. Given a graph G, a rumor set R and a budget B, we first feeds G as input to Algorithm 3, and obtains the processed graph G' in return. Next, we take the processed graph G', a rumor set R and a budget B as input to GreedySel and IndexSel. Note that we call these two pre-sampling based methods GreedySel* and IndexSel*, respectively.

Algorithm 3. PreSample(G)

3.1 **Input:** a graph G
3.2 **Output:** a reduced graph G'
   /* Sampling   */
3.3 Run r T-random walks for each node in $V \backslash R$
   /* reducing   */
3.4 **for** *each node* $v \in V \backslash R$ **do**
3.5 $\quad$ Compute P_v by scan $Is(R)$
3.6 $\quad$ **if** $P_v < 0.1$ **then**
3.7 $\quad\quad |$ $G' \leftarrow$ remove v from G
3.8 **return** G'

Complexity Analysis. In pre-sampling stage, Algorithm 3 takes $O(nrT)$ time to compute P_u for each node $u \in V \backslash R$. It is worth noting that we can't analyze how many nodes we can reduce in pre-sampling stage because it depends on the network structure and rumor set R. Therefore, the time complexity of GreedySel* and IndexSel* is $O(n(nL + r)T)$ and $O(n(L \log n + r)T)$, respectively. Although they theoretically have higher time complexity, they work very well in practice (see Sect. 5).

Algorithm 4. GreedySel*(G, R, B)

4.1 **Input:** a graph G, a rumor set R and a budget B
4.2 **Output:** a protector set P such that $c(P) < B$
   /* Pre-sampling   */
4.3 $G' \leftarrow$ PreSample(G)
   /* Greedy selection   */
4.4 $P \leftarrow$ GreedySel(G', R, B)
4.5 **return** P

Algorithm 5. IndexSel*(G, R, B)

5.1 **Input:** a graph G, a rumor set R and a budget B
5.2 **Output:** a protector set P such that $c(P) < B$
   /* Pre-sampling   */
5.3 $G' \leftarrow$ PreSample(G)
   /* Index based greedy selection   */
5.4 $P \leftarrow$ IndexSel(G', R, B)
5.5 **return** P

---

5   Experiments

In this section, we present our experiment results on effectiveness, efficiency, memory consumption and scalability of our proposed methods.

5.1   Experimental Settings

DataSets. We use three real-world datasets in the experiments: Gnutella, Email-Enron and Gowalla. All the datasets are obtained from an open-source website[2], and their statistics are shown in Table 2. The Gnutella dataset is a peer-to-peer file sharing network, the Email-Enron dataset is an email communication network, and the Gowalla dataset is a location-based social networking website where users share their locations by checking-in. The published graph data sets are unweighted, but since our method is not dependent on the semantics of the weights or their magnitude, we assign randomly generated weights (real numbers in the range 1 to 10) to the nodes of the graph.

Table 2. Summary of the datasets.

	n	m	#AvgDegree	#MaxDegree
Gnutella	8.8k	63k	7.2	88
Email-Enron	37k	184k	5.01	1383
Gowalla	197k	950k	4.83	14730

Algorithms. To the best of our knowledge, this is the first work to study the rumor containment within budget constraint on the random walk model, and thus there exists no previous work for direct comparison. In particular, we compare five methods, TopK, GreedySel (Algorithm 1), IndexSel (Algorithm 2), GreedySel* (Algorithm 4) and IndexSel* (Algorithm 5). It is worth noting that TopK is to select the top-k high unit block degree nodes as the targeted nodes.

Evaluation Metrics. We evaluate the performance of all methods by the runtime and the blocking percentage of the selected nodes. In particular, the percentage is computed by $\mathcal{G}(P|R)/Is(R)$, where $Is(R)$ denote the random walk set influenced by rumor set R. Moreover, to accurately measure $\mathcal{G}(P|R)$ for each algorithm, we compute it by running a sufficient number of MC simulations, i.e., $L = 1000$.

Parameter. Table 3 shows the settings of all parameters, such as the budget B, the size of the rumor set R, and the (random walk) length threshold T. To simulate the rumor set R, we select nodes uniformly at random from the nodes whose degrees are in top 10% of G.

---

[2] http://snap.stanford.edu/data/.

Table 3. Parameter setting.

Parameters	Values
B	50, 100, **150**, 200, 250
$|R|$	50, 100, **150**, 200, 250
T	3, 6, **9**, 12, 15

Setup. All codes are implemented in Java, and experiments are conducted on a server with 2.1 GHz Intel Xeon 8 Core CPU and 32 GB memory running CentOS/6.8 OS.

5.2   Effectiveness Test

This section studies how the block degree is affected by varying the budget B, the size of rumor set R and the length threshold T of a random walk.

Varying the Budget B. The block degrees of all algorithms on Gnutella and Email-Enron by varying the B are shown in Fig. 2, and we have the following observations on both datasets. (1) GreedySel and IndexSel achieve the same block degree, while TopK has the worst performance. GreedySel and IndexSel are little better than GreedySel* and IndexSel*. (2) With the growth of B, the advantage of GreedySel and IndexSel over TopK decreases, from 49% to 20% when B varies from 50 to 250 on Email-Enron dataset. This is because that TopK ignores the blocking overlaps between the selected seeds, which declines its block degree. However, when B is large, the performance of TopK catches up with the others since the overlaps are inevitable.

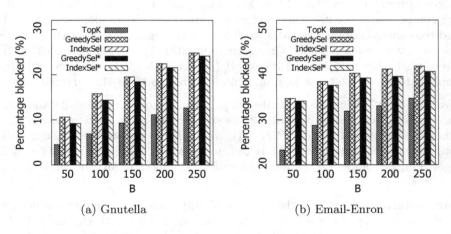

(a) Gnutella   (b) Email-Enron

Fig. 2. Effectiveness of varying the budget B

Varying the Size of R. Figure 3 shows the result by varying the size of R. We find: (1) with the growth of $|R|$, the blocking percentages of all methods are increasing because the increasing influence of R leads to more nodes with higher unit block degree. (2) GreedySel and IndexSel are consistently better than that of TopK, GreedySel* and IndexSel*.

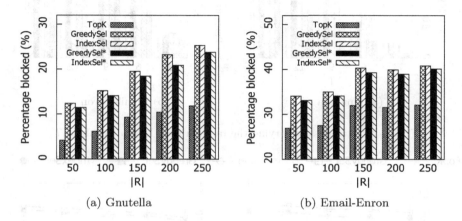

(a) Gnutella   (b) Email-Enron

Fig. 3. Effectiveness of varying the size of R

Varying the Random Walk Length Threshold T. Figure 4 shows the results by varying the threshold T, which determines the length of a random walk starting from a node. We observe that: (1) with the increase of T, the performance of all algorithms becomes better. The reason is that when the length becomes large, the random walk has more chances to reach the protectors and thus leads to a high unit block degree of the seeds. (2) The rumors on Gnutella dataset are much harder to be controlled than Email-Enron dataset. It implies that the network structure is an important variable for RCBC.

5.3 Efficiency Test

We evaluated the efficiency of different algorithms on Gnutella and Email-Enron datasets.

Varying the Budget B. Figure 5 presents the efficiency result when B varies from 50 to 250. We have the following observations. (1) The performance of IndexSel is about two orders of magnitude faster than GreedySel, which is consistent with our time complexity analysis. (2) The runtime of all methods except TopK is slowly increasing with the growth of B. This is because that the increasing of B directly causes selecting more nodes to P, which leads to an increase in the number of updating the block degree of the remaining node.

Varying the Size of R. Figure 6 shows the runtime of all algorithms on Gnutella and Email-Enron. We can see that the runtime of all methods except

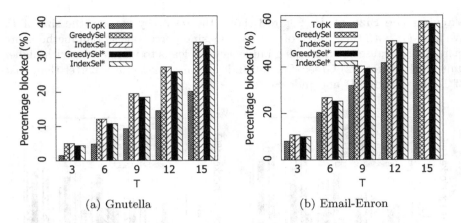

Fig. 4. Effectiveness of varying the random walk length threshold T

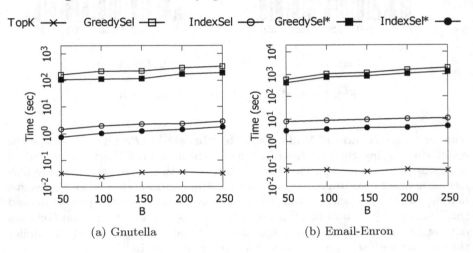

Fig. 5. Efficiency of varying the budget B

TopK is also slowly increasing when $|R|$ varies from 50 to 250 on all datasets. This is because the influence set $Is(R)$ of R is increasing with the growth of $|R|$.

Varying the Random Walk Length Threshold T. We evaluate the efficiencies of algorithms by varying T from 3 to 15. The result is shown in Fig. 7. We can see that all the algorithms except for TopK scale linearly with respect to T, which is consistent with our time complexity analysis.

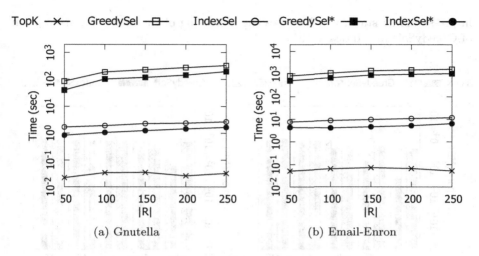

Fig. 6. Efficiency of varying the size of R

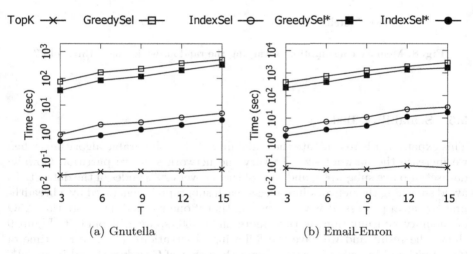

Fig. 7. Efficiency of varying the random walk length T

5.4   Memory Consumption

Figure 8 reports the average memory consumption of each algorithm when T is varying. As shown in Fig. 8, we find: (1) TopK, GreedySel and IndexSel have similar memory consumption. It is because that they all need to scan the sampling set to calculate the block degree of the nodes. (2) When T is small, the memory consumption of GreedySel* and IndexSel* is much smaller than that of other methods. This is because that only a small number of nodes can reach R when T is small. Therefore, we can reduce the sampling set by eliminating more nodes in pre-sampling stage. But with the growth of T, most nodes will reach R

in pre-sampling stage, which causes the increasing of the memory consumption of GreedySel* and IndexSel*.

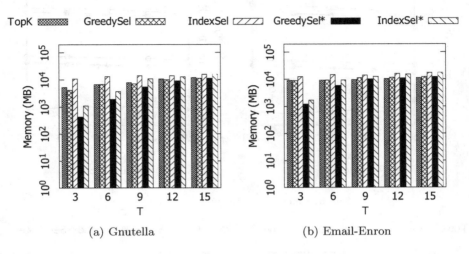

Fig. 8. Memory consumption of varying the random walk length threshold T

5.5 Scalability Test

This experiment is to evaluate the scalability of the comparable algorithms when we increase the network size. To vary the network size, we partition Gowalla dataset into five subgraphs, and each of them covers 20% nodes of the dataset. To avoid smashing the network into pieces, each subgraph is generated by a breadth-first traversal process. It is worth noting that if one method has more than 25G of memory consumption or runs more than 3000 s, we will omit it. Figure 9 shows the result and we have the following observations. (1) The run time of GreedySel and IndexSel increases faster than that of GreedySel* and IndexSel*, with the growth of graph size. This is because the runtime of GreedySel and IndexSelincreases with respect to n^2, while the runtime of GreedySel* and IndexSel*increases with respect to $n \log n$. (2) When the graph size is increasing, the memory consumption of TopK, GreedySel and IndexSel also increases faster than that of GreedySel* and IndexSel*. As we analyzed in Sect. 4.1, Pre-Sample can eliminate some node, which can't reach rumor set R. Therefore, PreSample can slow down the growth of the memory consumption when the graph size is increasing.

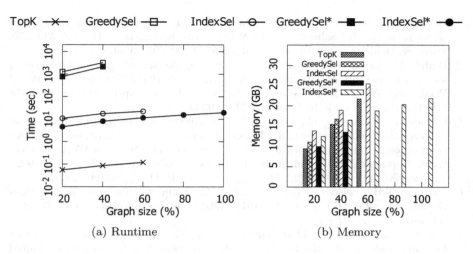

(a) Runtime

(b) Memory

Fig. 9. Scalability test on Gowalla dataset

6   Conclusion

In this paper, we proposed and studied the Rumor Control within Budget Constraint (RCBC) problem based on users' browsing behaviors (i.e., a random walk model), in order to actively limit the spread of rumors. We first proved the NP-hard of RCBC based on random walk model. Due to its hardness, we proposed two approximate algorithms (GreedySel and IndexSel) to solve this problem. To improve the efficiency of GreedySel and IndexSel, we proposed a PreSample method to eliminate the nodes that can't access the rumor set R by a T-length random walk. Lastly we conducted extensive experiments on real datasets to verify the efficiency, effectiveness, memory consumption and scalability of our methods.

Acknowledgements. This work is supported by the National Key Research and Development Program of China (Project Number: 2018YFB1003400) and the Fundamental Research Funds for the Central Universities (Project Number: 2042017kf1017).

References

1. Albert, R., Jeong, H., Barabási, A.L.: Error and attack tolerance of complex networks. Nature **406**(6794), 378 (2000)
2. Bharathi, S., Kempe, D., Salek, M.: Competitive influence maximization in social networks. In: Deng, X., Graham, F.C. (eds.) WINE 2007. LNCS, vol. 4858, pp. 306–311. Springer, Heidelberg (2007). https://doi.org/10.1007/978-3-540-77105-0_31

3. Borodin, A., Filmus, Y., Oren, J.: Threshold models for competitive influence in social networks. In: Saberi, A. (ed.) WINE 2010. LNCS, vol. 6484, pp. 539–550. Springer, Heidelberg (2010). https://doi.org/10.1007/978-3-642-17572-5_48
4. Budak, C., Agrawal, D., El Abbadi, A.: Limiting the spread of misinformation in social networks. In: Proceedings of the 20th International Conference on World Wide Web, WWW 2011, Hyderabad, India, 28 March–1 April 2011, pp. 665–674 (2011). https://doi.org/10.1145/1963405.1963499
5. Carnes, T., Nagarajan, C., Wild, S.M., van Zuylen, A.: Maximizing influence in a competitive social network: a follower's perspective. In: Proceedings of the 9th International Conference on Electronic Commerce: The Wireless World of Electronic Commerce, 2007, University of Minnesota, Minneapolis, MN, USA, 19–22 August 2007, pp. 351–360 (2007). https://doi.org/10.1145/1282100.1282167
6. Cormen, T.H., Leiserson, C.E., Rivest, R.L., Stein, C.: Introduction to Algorithms, 3rd edn. MIT Press, Cabridge (2009). http://mitpress.mit.edu/books/introduction-algorithms
7. Fan, L., Lu, Z., Wu, W., Thuraisingham, B.M., Ma, H., Bi, Y.: Least cost rumor blocking in social networks. In: IEEE 33rd International Conference on Distributed Computing Systems, ICDCS 2013, Philadelphia, Pennsylvania, USA, 8–11 July 2013, pp. 540–549 (2013). https://doi.org/10.1109/ICDCS.2013.34
8. Habiba, Yu, Y., Berger-Wolf, T.Y., Saia, J.: Finding spread blockers in dynamic networks. In: Giles, L., Smith, M., Yen, J., Zhang, H. (eds.) SNAKDD 2008. LNCS, pp. 55–76. Springer, Heidelberg (2008). https://doi.org/10.1007/978-3-642-14929-0_4
9. Hoeffding, W.: Probability inequalities for sums of bounded random variables. In: Fisher, N.I., Sen, P.K. (eds.) The Collected Works of Wassily Hoeffding, pp. 409–426. Springer, Heidelberg (1994). https://doi.org/10.1007/978-1-4612-0865-5_26
10. Khuller, S., Moss, A., Naor, J.: The budgeted maximum coverage problem. Inf. Process. Lett. **70**(1), 39–45 (1999). https://doi.org/10.1016/S0020-0190(99)00031-9
11. Nemhauser, G.L., Wolsey, L.A., Fisher, M.L.: An analysis of approximations for maximizing submodular set functions—I. Math. Program. **14**(1), 265–294 (1978). https://doi.org/10.1007/BF01588971
12. Nemhauser, G.L., Wolsey, L.A., Fisher, M.L.: An analysis of approximations for maximizing submodular set functions—i. Math. Program. **14**(1), 265–294 (1978)
13. Newman, M.E., Forrest, S., Balthrop, J.: Email networks and the spread of computer viruses. Phys. Rev. E **66**(3), 035101 (2002)
14. Nguyen, N.P., Yan, G., Thai, M.T., Eidenbenz, S.: Containment of misinformation spread in online social networks. In: Web Science 2012, WebSci 2012, Evanston, IL, USA, 22–24 June 2012, pp. 213–222 (2012). https://doi.org/10.1145/2380718.2380746
15. Spitzer, F.: Principles of Random Walk, vol. 34. Springer, Heidelberg (2013)
16. Tripathy, R.M., Bagchi, A., Mehta, S.: A study of rumor control strategies on social networks. In: Proceedings of the 19th ACM Conference on Information and Knowledge Management, CIKM 2010, Toronto, Ontario, Canada, 26–30 October 2010. pp. 1817–1820 (2010). https://doi.org/10.1145/1871437.1871737
17. Zhang, H., Zhang, H., Li, X., Thai, M.T.: Limiting the spread of misinformation while effectively raising awareness in social networks. In: Thai, M.T., Nguyen, N.P., Shen, H. (eds.) CSoNet 2015. LNCS, vol. 9197, pp. 35–47. Springer, Cham (2015). https://doi.org/10.1007/978-3-319-21786-4_4

18. Zhang, P., Bao, Z., Li, Y., Li, G., Zhang, Y., Peng, Z.: Trajectory-driven influential billboard placement. In: Proceedings of the 24th ACM SIGKDD International Conference on Knowledge Discovery & Data Mining, KDD 2018, London, UK, 19–23 August 2018, pp. 2748–2757 (2018). https://doi.org/10.1145/3219819.3219946
19. Zhang, P., et al.: Proactive rumor control in online networks. World Wide Web 1–20 (2018)

Quantum Reversible Fuzzy Grammars

Jianhua Jin[1,2(✉)] and Chunquan Li[2]

[1] School of Computer Science, Shaanxi Normal University, Xi'an 710119, China
jjh2006ok@aliyun.com
[2] School of Science, Southwest Petroleum University, Chengdu 610500, China
lichunquan@swpu.edu.cn

Abstract. Reversible languages are a class of regular languages that stands at the junction of several domains. While quantum reversible regular grammars are a special class of fuzzy regular grammars, the algebraic characterizations of quantum reversible regular languages have not been studied so far. This paper establishes quantum reversible fuzzy grammars and fuzzy (hyper-) regular grammars, and presents that the set of quantum reversible regular languages coincides with the set of all the quantum reversible hyper-regular languages. Moreover, it is shown that the set of quantum reversible regular languages coincides with that of quantum languages accepted by quantum reversible fuzzy automata. Particularly the algebraic properties of quantum reversible fuzzy grammars are discussed.

Keywords: Quantum fuzzy grammars · Fuzzy languages ·
Orthomodular lattice · Reversible fuzzy grammars · Automata

1 Introduction

As a mathematical model of quantum computation [15], automata theory based on quantum logic has been initiated by Ying in semantically analysis manner [21]. The quantum orthomodular lattice-valued predicate of recognizability and its fundamental properties have been discussed in detail, which shows the difference between classical computation and quantum computation in essence. Owing to the close relationship between automata theory and the theory of formal grammars, Cheng and Wang [6] established grammar theory based on quantum logic and in particular pointed out that the set of lattice-valued quantum regular languages generated by lattice-valued quantum regular grammars coincides with the set of lattice-valued quantum languages recognized by lattice-valued quantum automata. Actually, an orthomodular lattice is usually considered as quantum logic. Ying pointed out that the validity of Kleene theorem in

This work is supported by Fund of China Scholarship Council (No. 201708515152) and Graduate Educational Reform Project of Southwest Petroleum University (No.18YJZD08) and National Natural Science Foundation of China (Grant No. 11401495).

© Springer Nature Singapore Pte Ltd. 2019
X. Sun et al. (Eds.): NCTCS 2019, CCIS 1069, pp. 150–167, 2019.
https://doi.org/10.1007/978-981-15-0105-0_10

quantum computing was strongly dependent on the commutators of the truth-value domain orthomodular lattice [22]. Thereafter, by introducing the generalized subset construction method, Li has provided the fact that deterministic finite automata based on quantum logic is equivalent to quantum nondeterministic finite automata and thus completed Kleene theorem in quantum computing [10].

Reversible languages are a class of regular languages that stands at the junction of several domains. In 1973, Bennett proved the existence of reversible Turing machines [4], which could efficiently simulate classical Turing machines. The characterizations for reversible languages have been given by Lombardy [11]. Axelsen and Glück [3] discussed a universal reversible Turing machine from a programming language viewpoint, where programs were considered as part of both input and output of a universal reversible Turing machine. Recently Li and Li [9] introduced a reversible fuzzy automaton with membership values in the unit interval [0,1] and some characterizations of the languages accepted by reversible fuzzy automata. Reversible computing principles have been successfully applied in fields such as inversion [19], reversible programming [23] and translation [2].

Axelsen provided clean translation algorithms for generic reversible control flow operators, which avoided code duplication of the computation and thus lowered the power consumption of computing machinery [2]. As is well known, energy proves to be a pivotal computational resource, whose consumption in computation is deeply related to the reversibility of computation. However, it is necessary that no energy is dissipated in reversible computation. Quantum computing is unitary and thus reversible [18]. From this viewpoint, investigating the reversibility of computation is of great significance in quantum computing. Moreover, quantum computing is becoming a reality from not only the theoretical viewpoints but also the real-world applications. Large-scale quantum computers would theoretically be able to solve certain problems much more quickly than any classical computers that use even the best currently known algorithms, like integer factorization using Shor's algorithm [16]. New techniques are being discussed and carried out, making quantum operations feasible at room temperature [20]. New formal models also need to be proposed for future quantum machines. The contribution of this paper is mainly to discuss quantum reversible fuzzy grammars based on orthomodular lattices and study the algebraic properties of fuzzy languages they generated.

In recent years, there is an increasing interest for using fuzzy relational grammars in research fields such as handwritten mathematics and petroglyph recognition [8,12]. The fuzzy relational grammar, which not only involves the structure of the recognition domain but also considers the inherent uncertainty in recognizing that structure, is indeed suitable to model the recognition process. According to Chomsky classification, there are four types of fuzzy grammars including fuzzy regular, context-free, context sensitive and unrestricted grammars [13]. Chaudhari and Komejwar [5] stated that fuzzy regular grammars, fuzzy finite automata, fuzzy left linear grammars, fuzzy right linear grammars and fuzzy grammars in normal form are all equivalent in the sense that they generate the same languages, the languages of which are crisp languages here. Asveld investigated generalized fuzzy context-free grammars with truth-value domain in a completely distributive complete lattice [1]. The algebraic closure properties of the family of fuzzy context-free K-grammars languages were discussed in detail, which were very similar to those of crisp context-free languages. Recently, by replacing t-norm operator for several aggregation functions, fuzzy recursively enumerable languages and fuzzy recursive languages are established [17]. In particular, the class of fuzzy recursively enumerable languages prove to be closed under operations such as unions, widespread intersections, reverse and dual. Two normal forms for fuzzy linear grammars are introduced by Costa and Bedregal [7], the characterizations of which show that it is equivalent to both fuzzy linear automata and fuzzy 2-tape automata in the sense that they accept the same languages. Because the proposed quantum reversible fuzzy regular grammars are a special subclass of fuzzy regular grammars, the algebraic characterizations of quantum reversible regular languages will be investigated in this paper. As can be seen in the references [5,7,13], the normal form for quantum reversible fuzzy regular grammars will be also given. There is an open problem with regard to the algebraic closure properties of the family of quantum reversible regular languages. Hence, whether the family would be closed under algebraic operations such as union, intersection and reverse deserves careful attention.

This paper is arranged as follows. Section 2 presents the definitions of a quantum reversible fuzzy grammar, a quantum reversible fuzzy (hyper-) regular grammar and a quantum reversible fuzzy automaton. Then the equivalent relationship among quantum reversible fuzzy regular grammars, quantum reversible hyperregular grammars and quantum reversible fuzzy automata is discussed in the sense that they recognize the same classes of fuzzy languages. The algebraic characterizations of the quantum reversible fuzzy regular grammars are given. Section 3 investigates operations between quantum reversible regular languages and explores its algebraic closure properties. Some illustrative examples are also given to illustrate the construction method. Finally Sect. 4 gives the conclusions and some future work.

2   Quantum Reversible Fuzzy Grammars

Quantum logic is well known as a logic whose truth values are taken from an orthomodular lattice. An orthomodular lattice is a seven-tuple $(L, \leq, \vee, \wedge, \perp, 0, 1)$, where L is a nonempty set, $\leq$ is the partial ordering on L, the binary operations $\vee$ and $\wedge$ on L mean that $a \vee b \in L, a \wedge b \in L, \forall a, b \in L$, where $a \vee b$ and $a \wedge b$ stand for the least upper bound and the greatest lower bound of a and b respectively, the least element and the most element in L are denoted by 0 and 1 respectively, $\perp$ is a unary operation on L, which satisfies the following conditions: for any $a, b \in L, a \vee a^{\perp} = 1, a \wedge a^{\perp} = 0$; $a^{\perp\perp} = a$; $a \leq b$ implies $b^{\perp} \leq a^{\perp}$ and $a \vee (a^{\perp} \wedge b) = b$. If no confuse arises, we will abbreviate an orthomodular lattice $(L, \leq, \vee, \wedge, \perp, 0, 1)$ as the symbol L in this paper.

Definition 1. *Let* $(L, \leq, \vee, \wedge, \perp, 0, 1)$ *be an orthormodular lattice. A quantum fuzzy automaton is a 5-tuple* $\mathcal{M} = (Q, \Sigma, \delta, I, F)$, *where* Q *and* Σ *are finite nonempty sets of states and input symbols respectively;* $\delta : Q \times \Sigma \times Q \to L$ *is an* $L-$*fuzzy subset of* $Q \times \Sigma \times Q$, *called a quantum transition relation. Intuitively, for any* $p, q \in Q, x \in \Sigma$, $\delta(p, x, q)$ *expresses the truth value of the transition that inputting* x *makes state* p *transfer to state* q. $I, F : Q \to L$ *are* $L-$*fuzzy subsets of* Q, *called a fuzzy initial state and a fuzzy final state respectively.*

Let Σ^* be the free monoid generated from Σ with the operator of concatenation, where the empty string ε is identified with the identity of Σ. $\Sigma^+ = \Sigma \setminus \{\varepsilon\}$. Let

$$l(\Sigma^*) = \{f : \Sigma^* \to L | \ f \text{ is a mapping from } \Sigma^* \text{ to } L\}.$$

$l(\Sigma^*)$ denotes the set of all the fuzzy languages over Σ^*. In order to compute with words, the extension of δ, denoted by the mapping $\delta^* : Q \times \Sigma^* \times Q \to L$, is defined as follows: for any $q_0, q_n \in Q$, if $q_0 = q_n$, then $\delta^*(q_0, \varepsilon, q_n) = 1$; otherwise, $\delta^*(q_0, \varepsilon, q_n) = 0$; and for any $\theta = a_1 \cdots a_n \in \Sigma^*$ with $n \geq 1$, $\delta^*(q_0, \theta, q_n) = \vee\{\delta(q_0, a_1, q_1) \wedge \cdots \wedge \delta(q_{n-1}, a_n, q_n) | q_1, \ldots, q_{n-1} \in Q\}$.

For a quantum fuzzy automaton $\mathcal{M} = (Q, \Sigma, \delta, I, F)$, the language recognized by $\mathcal{M}$, denoted by $f_{\mathcal{M}}$, is defined as $f_{\mathcal{M}} \in l(\Sigma^*)$,

$$f_{\mathcal{M}}(\varepsilon) = \vee\{I(q) \wedge F(q) | q \in Q\},$$

and

$$f_{\mathcal{M}}(\theta) = \vee\{I(q_0) \wedge \delta(q_0, a_1, q_1) \wedge \cdots \wedge \delta(q_{n-1}, a_n, q_n) \wedge F(q_n) | q_0, \ldots, q_n \in Q\},$$

$\forall \theta = a_1 \cdots a_n \in \Sigma^+$, $a_i \in \Sigma$, $i = 1, \ldots, n$.

Definition 2. *Let* L *be an orthomodular lattice. Then a quantum fuzzy automaton* $\mathcal{M} = (Q, \Sigma, \delta, I, F)$ *over* L *is called reversible, if it satisfies the following conditions:*

(1) if $\delta(q, a, q_1) = \delta(q, a, q_2) > 0$, *then* $q_1 = q_2$, $\forall q_1, q \in Q, a \in \Sigma$;
(2) if $\delta(q_1, a, p) = \delta(q_2, a, p) > 0$, *then* $q_1 = q_2$, $\forall q_1, p \in Q, a \in \Sigma$.

Noting that a quantum reversible fuzzy automaton over an orthomodular lattice L will be abbreviated as $LQRFA$. The family of languages over Σ^*, accepted by all the $LQRFAs$, is denoted as $l_{LQRFAs}(\Sigma^*)$.

Definition 3. *A quantum fuzzy grammar is a system* $G = (N, T, P, I)$ *over an orthomodular lattice* L, *where* N *is a finite set of variables,* T *is a finite set of terminals, and* $T \cap N = \phi, I : N \to L$ *is a mapping named as a fuzzy initial variable,* P *is a finite collection of productions over* $T \cup N$, *and*

$$P = \{x \to y | x \in N^+, y \in (N \cup T)^*, \rho(x \to y) \in L \setminus \{0\}\},$$

where $\rho(x \to y)$ *represents the membership degree of the production* $x \to y$.

Denote $P_\varepsilon = \{x \to \varepsilon \mid x \in N^+, \rho(x \to \varepsilon) \in L \setminus \{0\}\}$. *If any* $x \to y_1 \in P \setminus P_\varepsilon$ *and* $x \to y_2 \in P \setminus P_\varepsilon$ *with* $\rho(x \to y_1) = \rho(x \to y_2)$ *imply that* $y_1 = y_2$, *and any* $x_1 \to y, x_2 \to y \in P \setminus P_\varepsilon$ *with* $\rho(x_1 \to y) = \rho(x_2 \to y)$ *implies that* $x_1 = x_2$, *then the system is called a quantum reversible fuzzy grammar.*

If β is derivable from $\alpha, i.e., \alpha \Rightarrow \beta$ by some production $x \rightarrow y$ in P, then we define

$$\rho(\alpha \Rightarrow \beta) = \rho(x \rightarrow y) \in L.$$

If $\alpha \Rightarrow^* \beta, i.e.$, either $\alpha = \beta$, then $\rho(\alpha \Rightarrow^* \beta) \in L$ or there exists $\alpha_1, \alpha_2, \cdots, \alpha_n$ (with $n \geq 2$) in N^* such that $\alpha = \alpha_1, \beta = \alpha_n$, with the corresponding production $x_i \rightarrow y_i$ for every $\alpha_i \Rightarrow \alpha_{i+1}, (i = 1, 2, \cdots, n-1)$, then

$$\rho(\alpha \Rightarrow \beta) = \bigwedge_{i=1}^{n-1} \rho(\omega_i \Rightarrow \omega_{i+1}) = \bigwedge_{i=1}^{n-1} \rho(\alpha_i \rightarrow \alpha_{i+1}).$$

A quantum reversible fuzzy grammar G is regular if it has only productions of the form:

$x \rightarrow \alpha y (\alpha \in T^+, x, y \in N)$ with $\rho(x \rightarrow \alpha y) \in L$

or of the form:

$x \rightarrow \alpha (\alpha \in T^*, x \in N)$ with $\rho(x \rightarrow \alpha) \in L$.

A quantum reversible fuzzy grammar G is hyper-regular if it has only productions of the form $x \rightarrow \alpha y (\alpha \in T, x, y \in N)$ with $\rho(x \rightarrow \alpha y) \in L$ or of the form $x \rightarrow \varepsilon (x \in N)$ with $\rho(x \rightarrow \varepsilon) \in L$

A quantum reversible fuzzy grammar generates a fuzzy language $f_G : T^* \rightarrow L, i.e.$, for any $\theta = w_n \in T^*, n \geq 1$, $f_G(\theta) = \bigvee \{I(\omega_0) \wedge \rho(\omega_0 \Rightarrow \omega_1) \wedge \cdots \wedge \rho(\omega_{n-1} \Rightarrow \omega_n) | \omega_0 \in N, \omega_1, \cdots, \omega_{n-1} \in (N \cup T)^*\}$

Noting that a quantum reversible (hyper-) regular language is one generated by some quantum reversible fuzzy (hyper-) regular grammar. $l_{QRRGs}(T^*)$ and $l_{QRHRGs}(T^*)$ will be represented as the family of quantum reversible regular languages and the family of quantum reversible hyper-regular languages over T^*, respectively.

Next we explore the relationship between the family of all the quantum reversible regular languages and the family of all the quantum reversible hyper-regular languages.

Proposition 1. $l_{QRRGs}(T^*) = l_{QRHRGs}(T^*)$.

Proof. From Definitions of quantum reversible fuzzy regular and hyper-regular grammars, every $L-$language generated by a quantum reversible fuzzy hyper-regular grammar can be clearly generated by a quantum reversible fuzzy regular grammar, which is the same as the former quantum reversible fuzzy hyper-regular grammar. That is, $l_{QRHRGs}(T^*) \subseteq l_{QRRGs}(T^*)$.

Next, it suffices to show that every $L-$language generated by a quantum reversible fuzzy regular grammar can be clearly generated by a quantum reversible fuzzy hyper-regular grammar. Let $G_1 = (N_1, T, I_1, P_1)$ be a quantum reversible fuzzy regular grammar over an orthomodular lattice L. We will define a a quantum reversible fuzzy hyper-regular grammar $G = (N, T, I, P)$ such that $f_G = f_{G_1}$, where $N_1 \subseteq N$.

(i) For each production

$$x \rightarrow a_1 \cdots a_m y \in P_1, \tag{1}$$

where $a_i \in T$ for $i = 1, \ldots, m$ and $x, y \in N_1$.

If $m = 1$, then $a_1 \in T, x, y \in N_1 \subseteq N$ and $x \to a_1 y \in P$ as required. If $m \geq 2$, then we define new nonterminal symbols $\xi_1, \cdots, \xi_{m-1}$ in N and within P mimic the production (1) by means of productions

$$x \to a_1 \xi_1, \cdots, \xi_{m-1} \to a_m y \in P$$

with

$$\rho_G(x \to a_1 \xi_1) = \cdots = \rho_G(\xi_{m-1} \to a_m y) = \rho_{G_1}(x \to a_1 \cdots a_m y), \qquad (2)$$

where ρ_G and ρ_{G_1} mean the membership degree of productions in G and G_1 respectively.

(ii) For each production

$$x \to b_1 \cdots b_n \in P_1, \qquad (3)$$

if $n = 1$ and $b_1 = \varepsilon$, then $x \in N_1 \subseteq N$, $x \to \varepsilon \in P$ and let $\rho_G(x \to \varepsilon) = \rho_{G_1}(x \to \varepsilon) \in L \setminus \{0\}$, as required. If $n \geq 1$, $b_i \in T$ for $i = 1, \ldots, n$ and $x \in N_1$, then we can define new nonterminal symbols $\eta_1, \ldots, \eta_n \in N$ and within P mimic the productions (3) by means of productions

$$x \to b_1 \eta_1, \cdots, \eta_{n-1} \to b_n \eta_n, \eta_n \to \varepsilon \in P$$

with

$$\rho_G(x \to b_1 \eta_1) = \cdots = \rho_G(\eta_{n-1} \to b_n \eta_n) = \rho_G(\eta_n \to \varepsilon) = \rho_{G_1}(x \to b_1 \cdots b_n). \qquad (4)$$

(iii) Put $N = N_1 \cup N_0$, where N_0 denotes the set of these new nonterminal symbols generated by productions (i) and (ii). Define the mapping $I : N \to L$ as follows: if $x \in N_1$, then $I(x) = I_1(x)$; otherwise, $I(x) = 0$.

Clearly G thus specified is a quantum reversible fuzzy hyper-regular grammar. Moreover, for any $x \in N_1$, we obtain $x \Rightarrow^* a_1 \cdots a_m y$ in G from the productions (2) with
$\rho_G(x \Rightarrow^* a_1 \cdots a_m y) = \rho_G(x \to a_1 \xi_1) \wedge \cdots \wedge \rho_G(\xi_{m-1} \to a_m y) = \rho_{G_1}(x \Rightarrow^* a_1 \cdots a_m y) \in L \setminus \{0\}$,
and
$x \Rightarrow^* b_1 \cdots b_n$ in G from the production (4) with
$\rho_G(x \Rightarrow^* b_1 \cdots b_n) = \rho_G(x \to b_1 \eta_1) \wedge \cdots \wedge \rho_G(\eta_{n-1} \to b_n \eta_n) \wedge \rho_G(\eta_n \to \varepsilon) = \rho_{G_1}(x \Rightarrow^* b_1 \cdots b_n) \in L \setminus \{0\}$.

Thus every derivation $x \Rightarrow^* \theta$ in G_1 with $\rho_{G_1}(x \Rightarrow^* \theta) \in L \setminus \{0\}$ can be simulated by a longer derivation $x \Rightarrow^* \theta$ in G with $\rho_G(x \Rightarrow^* \theta) = \rho_{G_1}(x \Rightarrow^* \theta) \in L \setminus \{0\}$. It follows that $f_{G_1} \leq f_G$.

To prove the reverse inequality, suppose that for any derivation $x \Rightarrow^* \theta$ in G with $\rho_G(x \Rightarrow^* \theta) \in L \setminus \{0\}$. Then certainly there is a derivation
$x \Rightarrow^* \theta$ in G_2 with

$$\rho_{G_2}(x \Rightarrow^* \theta) = \rho_G(x \Rightarrow^* \theta) \in L \setminus \{0\}, \qquad (5)$$

where $G_2 = (N, T, I, P \cup P_1)$ has all the productions in G together with all the productions in G_1.

Let $P \cup P_1 = P_2$. We will show that there is a derivation of θ in G_1 by induction on the number of symbols from N_0 which appears in the derivation (5). If no such symbols appear, then (5) is already a derivation in G_1. Otherwise the first appearance of a symbol of N_0 is based either on a production $x \to a_1 \xi_1$ with $\rho_{G_2}(x \to a_1\xi_1) = \rho_G(x \to a_1\xi_1) \in L \setminus \{0\}$, where $x \to a_1 \cdots a_m y$ is a production in G_1 or on a production $x \to b_1\eta_1$ with $\rho_{G_2}(x \to b_1\eta_1) = \rho_G(x \to b_1\eta_1) \in L \setminus \{0\}$, where $x \to b_1 \cdots b_n$ is a production in G_1. Consider the first of these cases. Since the final word θ in the derivation (5) has no nonterminal symbols and the grammar G_2 has no production of the type $\xi_i \to \theta$ ($\theta \in T^*$) with $\rho_{G_2}(\xi_i \to \theta) \in L \setminus \{0\}$ for any symbol of N_0, the only way in which ξ_1, once introduced, can subsequently disappear must involve changes from ξ_1 to $a_2\xi_2$, $\cdots, \xi_{m-1}$ to $a_m y$.

However, the sequence of productions $x \to a_1\xi_1, \xi_1 \to a_2\xi_2, \cdots, \xi_{m-1} \to a_m y$ with $\rho_{G_2}(x \to a_1\xi_1) = \rho_G(x \to a_1\xi_1)$, $\rho_{G_2}(\xi_1 \to a_2\xi_2) = \rho_G(\xi_1 \to a_2\xi_2)$, $\cdots$, $\rho_{G_2}(\xi_{m-1} \to a_m y) = \rho_G(\xi_{m-1} \to a_m y) \in L \setminus \{0\}$ can be replaced by a single transition $x \to a_1 a_2 \cdots a_m y$ in G_2 with

$$\rho_{G_2}(x \to a_1 a_2 \cdots a_m y) = \rho_{G_2}(x \to a_1\xi_1) \wedge \rho_{G_2}(\xi_1 \to a_2\xi_2) \wedge \cdots \wedge \rho_{G_2}(\xi_{m-1} \to a_m y) = \rho_G(x \to a_1\xi_1) \wedge \rho_G(\xi_1 \to a_2\xi_2) \wedge \cdots \wedge \rho_G(\xi_{m-1} \to a_m y) = \rho_{G_1}(x \to a_1 a_2 \cdots a_m y).$$

Thus the number of symbols from N_0 has been reduced.

Equally, in the second case, the derivation must involve subsequent changes from η_1 to $b_2\eta_2$, $\cdots$, η_{n-1} to $b_n\eta_n$, and these transitions can be replaced by a single transition in G_2 from x to $b_1 b_2 \cdots b_n$ with

$$\rho_{G_2}(x \to b_1 b_2 \cdots b_n) = \rho_{G_2}(x \to b_1\eta_1) \wedge \rho_{G_2}(\eta_1 \to b_2\eta_2) \wedge \cdots \wedge \rho_{G_2}(\eta_n \to \varepsilon) = \rho_G(x \to b_1\eta_1) \wedge \rho_G(\eta_1 \to b_2\eta_2) \wedge \cdots \wedge \rho_G(\eta_n \to \varepsilon) = \rho_{G_1}(x \to b_1 b_2 \cdots b_n).$$

In both cases the derivation (5) is replaced by one with fewer occurrences of symbols from N_0 with $f_{G_2}(\theta) = f_{G_1}(\theta)$. By induction it now follows that $f_{G_2} \leq f_{G_1}$. Hence, certainly $f_G \leq f_{G_1}$.

Therefore, we have $f_G = f_{G_1}$. Hence $l_{QRRGs}(T^*) \subseteq l_{QRHRGs}(T^*)$.

From the above, this completes the proof that $l_{QRRGs}(T^*) = l_{QRHRGs}(T^*)$.

Remark 1. It sees from Proposition 1 that quantum reversible fuzzy regular grammars and quantum reversible fuzzy hyper-regular grammars are equivalent in the sense that they generate the same family of fuzzy languages. Owing to the simpler production expressions in quantum reversible fuzzy hyper-regular grammars, a quantum reversible fuzzy hyper-regular grammar would be called as the normal form of quantum reversible fuzzy regular grammar.

Next we investigate the relationship between the family of all the quantum reversible regular languages and the family of languages accepted by $LQRFAs$.

Theorem 1. $l_{QRRGs}(T^*) = l_{LQRFAs}(T^*)$.

Proof. Claim 1. $l_{QRRGs}(T^*) \subseteq l_{LQRFAs}(T^*)$.

Let $G = (N, T, I, P)$ be a quantum reversible fuzzy regular grammar over an orthomodular lattice L. Then there exists a quantum reversible fuzzy automaton $\mathcal{M}$ over L such that $f_{\mathcal{M}} = f_G$. In fact, construct the quantum fuzzy automaton $\mathcal{M} = (N, T, \delta, I, F)$, where N, T and I are the same as those in G,

$$\delta(q, a, p) = \begin{cases} \rho(q \to ap), & \text{if } q \to ap \in P \\ 0, & \text{otherwise} \end{cases}$$

and

$$F(q) = \begin{cases} \rho(q \to \varepsilon), & \text{if } q \to \varepsilon \in P \\ 0, & \text{otherwise.} \end{cases}$$

Clearly $\mathcal{M} = (N, T, \delta, I, F)$ constructed above is a quantum reversible fuzzy automaton.

For any $\theta \in T^*$, if $\theta = \varepsilon$, then

$f_{\mathcal{M}}(\theta) = \vee\{I(q) \wedge \delta^*(q, \varepsilon, q) \wedge F(q) | q \in N\} = \vee\{I(q) \wedge F(q) | q \in N\} = \vee\{I(q) \wedge \rho(q \Rightarrow \varepsilon) | q \in N\} = f_G(\theta)$.

If $\theta \neq \varepsilon$, then suppose $\theta = a_1 \cdots a_n, a_i \in T$ for $i = 1, \ldots, n$.

If $f_G(\theta) \in L \setminus \{0\}$, then it follows that

$f_{\mathcal{M}}(\theta) = \vee\{I(q) \wedge \delta(q, a_1, q_1) \wedge \cdots \wedge \delta(q_{n-1}, a_n, q_n) \wedge \rho(q_n \to \varepsilon) | q, q_1, \ldots, q_n \in N\} = \vee\{I(q) \wedge \rho(q \to a_1 q_1) \wedge \cdots \wedge \rho(q_{n-1} \to a_n q_n) \wedge \rho(q_n \to \varepsilon) | q, q_1, \ldots, q_n \in N\} = f_G(\theta)$. So $f_G = f_{\mathcal{M}}$.

Hence $l_{QRRGs}(T^*) \subseteq l_{LQRFAs}(T^*)$.

Claim 2. $l_{LQRFAs}(T^*) \subseteq l_{QRRGs}(T^*)$.

Let $\mathcal{M} = (Q, T, \delta, I, F)$ be a quantum reversible fuzzy automaton over an orthomodular lattice L. Then there exists a quantum reversible fuzzy regular grammar $G = (Q, T, I, P)$ such that $f_G = f_M$. In fact, We define a quantum fuzzy regular grammar as follows: $G = (Q, T, I, P)$, where P consists of the productions $p \to aq(p, q \in Q, a \in T)$ with $\rho_G(p \to aq) = \delta(p, a, q) \in L \setminus \{0\}$, and $q \to \varepsilon (q \in Q)$ with $\rho_G(q \to \varepsilon) = F(q) \in L \setminus \{0\}$.

Clearly $G = (Q, T, I, P)$ constructed above is a quantum reversible fuzzy regular grammar.

For any $\theta \in \Sigma^*$, if $\theta = \varepsilon$, then

$f_G(\varepsilon) = \vee\{I(q) \wedge \rho_G(q \to \varepsilon) | q \in Q\} = \vee\{I(q) \wedge F(q) | q \in Q\} = f_{\mathcal{M}}(\varepsilon)$;

if $\theta \in T^* \setminus \{\varepsilon\}$, then it follows that

$f_G(\theta) = \vee\{I(q) \wedge \rho(q \Rightarrow a_1 q_1) \wedge \rho(a_1 q_1 \Rightarrow a_1 a_2 q_2) \wedge \cdots \wedge \rho(a_1 a_2 \cdots a_{n-1} q_{n-1} \Rightarrow a_1 a_2 \cdots a_n q_n) \wedge \rho(a_1 a_2 \cdots a_n q_n \Rightarrow a_1 a_2 \cdots a_n) | q, q_1, \ldots q_n \in Q\} = \vee\{I(q) \wedge \rho(q \to a_1 q_1) \wedge \cdots \wedge \rho(q_n \to \varepsilon) | q \in Q, q_1, \ldots, q_n \in Q\} = \vee\{I(q) \wedge \delta(q, a_1, q_1) \wedge \cdots \wedge \delta(q_{n-1}, a_n, q_n) \wedge F(q_n)) | q, q_1, \ldots, q_n \in Q\} = f_{\mathcal{M}}(\theta)$.

So $f_{\mathcal{M}} = f_G$ and $l_{LQRFAs}(T^*) \subseteq l_{QRRGs}(T^*)$.

This completes the proof from the above that $l_{QRRGs}(T^*) = l_{LQRFAs}(T^*)$.

Example 1. Let a Hilbert space be denoted as $H = \bigotimes^2 C^2$ [14], where C denotes the set of complex numbers. Consider the set L_2 which is called all closed subspaces of H. If the inclusion relation of sets is denoted as partial order relation $\leq$, and $\wedge$ is defined as intersection of sets, $\vee$ is defined as the closure of union of sets, then clearly, $l_2 = (L_2, \leq, \wedge, \vee, 0, 1)$ is a complete lattice, where 0 and 1 are 0 (denotes 0 vector) and H respectively. Furthermore, if $A^\perp$ is defined as the orthocomplement of A, then l_2 is a complete orthomodular lattice.

As the standard notation in quantum computation [14], $\mid 0\rangle \mid 0\rangle, \mid 0\rangle \mid 1\rangle, \mid 1\rangle \mid 0\rangle$ and $\mid 1\rangle \mid 1\rangle$ are four computational basis states in the two-qubits state space H. Noting that $\mid 0\rangle \mid 0\rangle, \mid 0\rangle \mid 1\rangle, \mid 1\rangle \mid 0\rangle$ and $\mid 1\rangle \mid 1\rangle$ also represent an orthonormal base of H, where the set of elements $\mid 0\rangle$ and $\mid 1\rangle$ consists of an orthonormal base of space C^2. The subspace spanned by $\mid i\rangle \mid j\rangle$, denote by $a_{ij} = span\{\mid i\rangle \mid j\rangle\}, i, j \in \{0, 1\}$.

Let $G = (N, T, P, I)$ be a quantum fuzzy grammar over a lattice l_2, where $N = \{p, q\}, T = \{\sigma\}, I$ is a map from N to l_2, i.e., $I(p) = 1$, and $I(q) = a_{11}$, and P is composed of the following productions:

$p \to \sigma q$ with $\rho(p \to \sigma q) = a_{00}$;
$p \to \sigma p$ with $\rho(p \to \sigma p) = a_{01}$;
$q \to \sigma p$ with $\rho(q \to \sigma p) = a_{11}$;
$q \to \sigma q$ with $\rho(q \to \sigma q) = a_{10}$;
$p \to \varepsilon$ with $\rho(p \to \varepsilon) = a_{11}$;
$q \to \varepsilon$ with $\rho(q \to \varepsilon) = 1$.

$G = (N, T, P, I)$ is obviously a quantum reversible fuzzy grammar over a lattice l_2. The fuzzy language recognized by G could be obtained as follows:

$$f_G(\varepsilon) = a_{11},$$

$f_G(\sigma) = (I(p) \wedge \rho(p \to \sigma q) \wedge \rho(q \to \varepsilon)) \vee (I(p) \wedge \rho(p \to \sigma p) \wedge \rho(p \to \varepsilon)) \vee (I(q) \wedge \rho(q \to \sigma q) \wedge \rho(q \to \varepsilon)) \vee (I(q) \wedge \rho(q \to \sigma p) \wedge \rho(p \to \varepsilon)) \vee = a_{00} \vee a_{11}$,
and

$$f_G(\theta) = 0, \quad \theta \in T^* \setminus \{\varepsilon, \sigma\}.$$

According to Theorem 1, we could construct a quantum reversible fuzzy automaton $\mathcal{M} = (Q, \Sigma, \delta, I', F)$ over l_2 equivalent to the fuzzy grammar G, where $Q = N, \Sigma = T, I'$ is a map from Q to l_2, i.e., $I'(p) = 1, I'(q) = a_{11}, F$ is a map from Q to l_2, i.e., $F(p) = \rho(p \to \varepsilon) = a_{11}, F(q) = \rho(q \to \varepsilon) = 1$, and δ is a map from $Q \times \Sigma \times Q$ to l_2, i.e.,

$\delta(p, \sigma, q) = a_{00}$,
$\delta(p, \sigma, p) = a_{01}$,
$\delta(q, \sigma, p) = a_{11}$,
$\delta(q, \sigma, q) = a_{10}$.

We draw an $LQRFA$ as follows. The states are in circles, and particularly whenever $F(q) \neq 0$, the state q is marked by a blue circle. For state q with $I(q) \neq 0$, we draw into q an edge without a source labeled $start$. A transition $q \xrightarrow{\sigma, u} p$ is depicted as shown in Fig. 1, which means that the current state q is transformed into state p with membership degree of the transition denoted by $\delta(q, \sigma, p) = u$ whenever the alphabet σ occurs.

Clearly, $f_M = f_G$.

Example 1 shows that how to construct an equivalent quantum reversible fuzzy automaton when a quantum reversible fuzzy grammar is given. Conversely, if a quantum reversible fuzzy automaton is given, then an equivalent reversible fuzzy regular grammar can be easily established by Theorem 1.

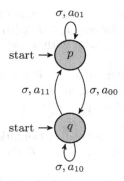

Fig. 1. Graphical representation of an $LQRFA$

Definition 4. *A λ-level grammar of fuzzy grammar $G = (N, T, P, I)$ over an orthomodular lattice is a classic grammar $G_{[\lambda]} = (N, T, P_{[\lambda]}, I_{[\lambda]})$, where $I_{[\lambda]}$ is the set of initial variables, and $\forall q \in N, q \in I_{[\lambda]}$ if and only if $I(q) = \lambda$, $P_{[\lambda]}$ is a finite set consisting of productions, i.e.,*

$$P_{[\lambda]} = \{x \to y | x \in N^+, y \in (N \cup T)^*, \rho(x \to y) = \lambda\}.$$

Definition 5. *Let f be a fuzzy language from T^* to L,*

$$f_{[\lambda]} = \{\theta \in T^* | f(\theta) = \lambda\}$$

is called λ-level set of f, where $\lambda \in L$.

$f_{[\lambda]}$ is a crisp language on T^*.

Theorem 2. *Let L be an orthomodular lattice. Then for any quantum fuzzy regular language $f : T^* \to L$,*

$$f = \bigvee_{\lambda \in L} \lambda \wedge f_{[\lambda]}.$$

Proof. For any $\theta \in T^*$,

$$f_{[\lambda]}(\theta) = \begin{cases} 1, & \theta \in f_{[\lambda]} \\ 0, & \theta \notin f_{[\lambda]} \end{cases}$$

$(\bigvee_{\lambda \in L} \lambda \wedge f_{[\lambda]})(\theta) = \bigvee_{\lambda \in L} \lambda \wedge f_{[\lambda]}(\theta) = \bigvee_{\theta \in f_{[\lambda]}} \lambda = \bigvee_{f(\theta) = \lambda} \lambda = f(\theta)$.

   Therefore,

$$f = \bigvee_{\lambda \in L} \lambda \wedge f_{[\lambda]}.$$

Proposition 2. *For a quantum fuzzy regular grammar $G = (N, T, P, I)$ over an orthomodular lattice L, G is reversible if and only if $\forall \lambda \in L \setminus \{0\}, G_{[\lambda]}$ is reversible.*

Proof. Suppose a quantum reversible fuzzy regular grammar be $G = (N, T, P, I)$ over L. Since G is reversible, for any $r \in L \setminus \{0\}, x_1 \to \alpha y \in P, x_2 \to \alpha y \in P, \rho(x_1 \to \alpha y) = r, \rho(x_2 \to \alpha y) = r$ with $x_1, x_2, y \in N, \alpha \in T^+$ imply that $x_1 = x_2$;

$\forall r \in L \setminus \{0\}, x \to \alpha y_1 \in P, x \to \alpha y_2 \in P, \rho(x \to \alpha y_1) = r, \rho(x \to \alpha y_2) = r$ with $x, y_1, y_2 \in N, \alpha \in T^+$ imply that $y_1 = y_2$;

$x_1 \to \alpha \in P, x_2 \to \alpha \in P, \rho(x_1 \to \alpha) = r, \rho(x_2 \to \alpha) = r$ with $x_1, x_2 \in N, \alpha \in T^+$ imply that $x_1 = x_2$.

Therefore, $\forall \lambda \in L \setminus \{0\}, G_{[\lambda]}$ defined as Definition 4 is reversible.

Conversely, if $\forall \lambda \in L \setminus \{0\}, G_{[\lambda]} = (N, T, P_{[\lambda]}, I_{[\lambda]})$ is reversible, then the following results are valid:

Whenever $x \to \alpha y_1 \in P_{[\lambda]}, x \to \alpha y_2 \in P_{[\lambda]}$ with any $x, y_1, y_2 \in N, \alpha \in T^+$, there must be $y_1 = y_2$.

Whenever $x_1 \to \alpha \in P_{[\lambda]}$ and $x_2 \to \alpha \in P_{[\lambda]}$ with any $x_1, x_2 \in N, \alpha \in T^+$, there must be $x_1 = x_2$.

Since

$$P = \bigcup_{\lambda \in L \setminus \{0\}} P_{[\lambda]},$$

it is easily concluded by Definition 4 that $G = (N, T, P, I)$ over L is a quantum reversible fuzzy regular grammar.

3   Algebraic Properties of Quantum Reversible Regular Languages

Next we will investigate algebraic operations of quantum regular languages. For any quantum regular languages $f, g : \Sigma^* \to L$, the union between f and g, denoted by $f \cup g$, is defined as $f \cup g(\omega) = f(\omega) \vee g(\omega), \forall \omega \in \Sigma^*$; the intersection between f and g, denoted by $f \cap g$, is defined as $f \cap g(\omega) = f(\omega) \wedge g(\omega), \forall \omega \in \Sigma^*$; and the reverse of f, denoted by f^{-1}, is given by $f^{-1}(\omega) = f(\omega^{-1}), \forall \omega \in \Sigma^*$, where ω^{-1} is the reverse of ω, i.e., if $\omega = \sigma_1 \sigma_2 \cdots \sigma_n$, then $\omega^{-1} = \sigma_n \sigma_{n-1} \cdots \sigma_1$, where $\sigma_i \in \Sigma, i = 1, \cdots, n$.

Proposition 3. *The family of quantum reversible regular languages is closed under union.*

Proof. Suppose that $G_1 = (N_1, T, P_1, I_1)$ and $G_2 = (N_2, T, P_2, I_2)$ are two quantum reversible fuzzy hyper-regular grammars over an orthomodular lattice L, and the languages they generated are f_1 and f_2 respectively.

Without loss of generality, we could assume that $N_1 \cap N_2 = \phi$. Construct a quantum fuzzy hyper-regular grammar $G = (N, T, P, I)$ over L, where $N = N_1 \cup N_2, P = P_1 \cup P_2$ and a mapping $I : N \to L$ is defined as follows: $I(q) = I_1(q)$ if $q \in N_1$, and $I(q) = I_2(q)$ if $q \in N_2$.

Claim 1. It follows that from the productions P that the above constructed grammar $G = (N, T, P, I)$ is reversible.

Claim 2. $G = (N, T, P, I)$ generates the union of quantum reversible regular languages f_1 and f_2.

Noting that $f_1 \cup f_2(\theta) = f_1(\theta) \vee f_2(\theta), \forall \theta \in T^*$. Let f_G represent the language generated by G. Then

$f_G(\varepsilon) = \bigvee\{I(q) \wedge \rho_G(q \rightarrow \varepsilon) \mid q \in N\} = (\bigvee\{I_1(q) \wedge \rho_{G_1}(q \rightarrow \varepsilon) \mid q \in N_1\}) \vee (\bigvee\{I_2(q) \wedge \rho_{G_2}(q \rightarrow \varepsilon) \mid q \in N_2\}) = f_1(\varepsilon) \vee f_2(\varepsilon) = f_1 \cup f_2(\varepsilon).$

For any $\theta = a_1 a_2 \cdots a_n \in T^+$ with $a_1, a_2, \cdots, a_n \in T$,

$f_G(\theta) = \bigvee\{I(q) \wedge \rho_G(q \rightarrow a_1 q_1) \wedge \cdots \wedge \rho_G(q_{n-2} \rightarrow a_{n-1} q_{n-1}) \wedge \rho_G(q_{n-1} \rightarrow a_n q_n) \wedge \rho_G(q_n \rightarrow \varepsilon) | q, q_1, \ldots, q_{n-1}, q_n \in N\} = (\bigvee\{I_1(q) \wedge \rho_{G_1}(q \rightarrow a_1 q_1) \wedge \cdots \wedge \rho_{G_1}(q_{n-2} \rightarrow a_{n-1} q_{n-1}) \wedge \rho_{G_1}(q_{n-1} \rightarrow a_n q_n) \wedge \rho_{G_1}(q_n \rightarrow \varepsilon) | q, q_1, \ldots, q_{n-1}, q_n \in N_1\}) \vee (\bigvee\{I_2(q) \wedge \rho_{G_2}(q \rightarrow a_1 q_1) \wedge \cdots \wedge \rho_{G_2}(q_{n-2} \rightarrow a_{n-1} q_{n-1}) \wedge \rho_{G_2}(q_{n-1} \rightarrow a_n q_n) \wedge \rho_{G_2}(q_n \rightarrow \varepsilon) | q, q_1, \ldots, q_{n-1}, q_n \in N_2\}) = f_1(\theta) \vee f_2(\theta) = f_1 \cup f_2(\theta).$

Hence, f_G is the union of f_1 and f_2. This completes the proof.

Proposition 4. *Let mappings $f_1 : \Sigma^* \rightarrow L_1$ and $f_2 : \Sigma^* \rightarrow L_2$ be two quantum reversible regular languages. Then a mapping $g : \Sigma^* \rightarrow L_1 \times L_2$, given by $g(\omega) = (f_1(\omega), f_2(\omega)), \forall \omega \in \Sigma^*$, is also a quantum reversible regular language.*

Proof. Suppose that quantum reversible fuzzy automata $\mathcal{A} = (Q_1, \Sigma, \delta_1, I_1, F_1)$ over L_1 and $\mathcal{B} = (Q_2, \Sigma, \delta_2, I_2, F_2)$ over L_2 accept quantum reversible regular languages f_1 and f_2, respectively. Construct a fuzzy automaton $\mathcal{M} = (Q, \Sigma, \delta, I, F)$ over $L_1 \times L_2$ as follows: $Q = Q_1 \times Q_2$, $I : Q \rightarrow L_1 \times L_2$ is a mapping given by $I((q_1, q_2)) = (I_1(q_1), I_2(q_2)), \forall (q_1, q_2) \in Q$; $F : Q \rightarrow L_1 \times L_2$ is a mapping given by $F((q_1, q_2)) = (F_1(q_1), F_2(q_2)), \forall (q_1, q_2) \in Q$; and $\delta : Q \times \Sigma \times Q \rightarrow L_1 \times L_2$ is a mapping defined by $\delta((q_1, q_2), a, (p_1, p_2)) = (\delta_1(q_1, a, p_1), \delta_2(q_2, a, p_2)), \forall (q_1, q_2), (p_1, p_2) \in Q, a \in \Sigma.$

Claim 1. $L_1 \times L_2$ is an orthomodular lattice.

For any $(c_1, c_2), (d_1, d_2) \in L_1 \times L_2$, $(c_1, c_2) \leq (d_1, d_2)$ is defined as $c_1 \leq d_1$ and $c_2 \leq d_2$, $(c_1, c_2) \vee (d_1, d_2) = (c_1 \vee d_1, c_2 \vee d_2)$, $(c_1, c_2) \wedge (d_1, d_2) = (c_1 \wedge d_1, c_2 \wedge d_2)$, $(c_1, c_2)^\perp = (c_1^\perp, c_2^\perp)$. Then $\leq$ is a partial ordering on $L_1 \times L_2$, $(0, 0)$ and $(1, 1)$ are the least and greatest element respectively in $L_1 \times L_2$, $(c_1, c_2)^{\perp\perp} = (c_1, c_2)$, $(c_1, c_2) \vee (c_1, c_2)^\perp = (1, 1)$ and $(c_1, c_2) \wedge (c_1, c_2)^\perp = (0, 0)$. If $(c_1, c_2) \leq (d_1, d_2)$, then $(d_1, d_2)^\perp \leq (c_1, c_2)^\perp$, $(c_1, c_2) \vee ((c_1, c_2)^\perp \wedge (d_1, d_2)) = (c_1, c_2) \vee ((c_1^\perp, c_2^\perp) \wedge (d_1, d_2)) = (c_1, c_2) \vee (c_1^\perp \wedge d_1, c_2^\perp \wedge d_2) = (c_1 \vee (c_1^\perp \wedge d_1), c_2 \vee (c_2^\perp \wedge d_2)) = (d_1, d_2)$. Hence, $(L_1 \times L_2, \leq, \vee, \wedge, \perp, (0, 0), (1, 1))$ is an orthomodular lattice.

Claim 2. $\mathcal{M} = (Q, \Sigma, \delta, I, F)$ over $L_1 \times L_2$ is reversible.

For any $(q_1, q_2), (p_1, p_2) \in Q, a \in \Sigma$, if

$$\delta((q_1, q_2), a, (p_1, p_2)) = \delta((q_1, q_2), a, (r_1, r_2)) \in L_1 \times L_2 \backslash \{(0, 0)\},$$

then $(\delta_1(q_1, a, p_1), \delta_2(q_2, a, p_2)) = (\delta_1(q_1, a, r_1), \delta_2(q_2, a, r_2))$. So $\delta_1(q_1, a, p_1) = \delta_1(q_1, a, r_1)$ and $\delta_2(q_2, a, p_2) = \delta_2(q_2, a, r_2)$. Since quantum fuzzy automata $\mathcal{A}$ and $\mathcal{B}$ are reversible, $p_1 = r_1$ and $p_2 = r_2$. Hence $(p_1, p_2) = (r_1, r_2)$. Similarly, it is easily shown that, if $\delta((q_1, q_2), a, (p_1, p_2)) = \delta((r_1, r_2), a, (p_1, p_2)) \in L_1 \times L_2 \backslash \{(0, 0)\}$ for any $(q_1, q_2), (p_1, p_2) \in Q, a \in \Sigma$, then $(q_1, q_2) = (r_1, r_2)$. By Definition 2, $\mathcal{M} = (Q, \Sigma, \delta, I, F)$ over $L_1 \times L_2$ is reversible.

Claim 3. $\mathcal{M} = (Q, \Sigma, \delta, I, F)$ over $L_1 \times L_2$ accepts the mapping $g : \Sigma^* \to L_1 \times L_2$ given by $g(\omega) = (f_1(\omega), f_2(\omega)), \forall \omega \in \Sigma^*$.

Let $f_{\mathcal{M}}$ be the language accepted by $\mathcal{M} = (Q, \Sigma, \delta, I, F)$ over $L_1 \times L_2$. Then
$f_{\mathcal{M}}(\varepsilon) = \bigvee\{I((q_1, q_2)) \wedge F((q_1, q_2)) \mid (q_1, q_2) \in Q\} = \bigvee\{(I_1(q_1), I_2(q_2)) \wedge (F_1(q_1), F_2(q_2)) \mid (q_1, q_2) \in Q\} = \bigvee\{(I_1(q_1) \wedge F_1(q_1), I_2(q_2) \wedge F_2(q_2)) \mid (q_1, q_2) \in Q\} = (\bigvee\{I_1(q_1) \wedge F_1(q_1) \mid q_1 \in Q_1\}, \bigvee\{I_2(q_2) \wedge F_2(q_2) \mid q_2 \in Q_2\}) = (f_1(\varepsilon), f_2(\varepsilon))$.

For any $\omega \in \Sigma^+$ with length $|\omega| = n$, let $\omega = a_1 \cdots a_n$, $a_i \in \Sigma, i = 1, \cdots, n$. Then
$f_{\mathcal{M}}(\omega) = \bigvee\{I((q_{01}, q_{02})) \wedge \delta((q_{01}, q_{02}), a_1, (q_{11}, q_{12})) \wedge \delta((q_{11}, q_{12}), a_2, (q_{21}, q_{22})) \wedge \cdots \wedge \delta((q_{n-1,1}, q_{n-1,2}), a_n, (q_{n1}, q_{n2})) \wedge F((q_{n1}, q_{n2})) \mid (q_{i1}, q_{i2}) \in Q, i = 0, 1, \ldots, n\} = \bigvee\{(I_1(q_{01}), I_2(q_{02})) \wedge (\delta_1(q_{01}, a_1, q_{11}), \delta_2(q_{02}, a_1, q_{12})) \wedge (\delta_1(q_{11}, a_2, q_{21}), \delta_2(q_{12}, a_2, q_{22})) \wedge \cdots \wedge (\delta_1(q_{n-1,1}, a_n, q_{n1}), \delta_2(q_{n-1,2}, a_n, q_{n2})) \wedge (F_1(q_{n1}), F_2(q_{n2})) \mid (q_{i1}, q_{i2}) \in Q, i = 0, 1, \ldots, n\} = \bigvee\{(I_1(q_{01}) \wedge \delta_1(q_{01}, a_1, q_{11}) \wedge \delta_1(q_{11}, a_2, q_{21}) \wedge \cdots \wedge \delta_1(q_{n-1,1}, a_n, q_{n1}) \wedge F_1(q_{n1}), I_2(q_{02}) \wedge \delta_2(q_{02}, a_1, q_{12}) \wedge \delta_2(q_{12}, a_2, q_{22}) \wedge \cdots \wedge \delta_2(q_{n-1,2}, a_n, q_{n2}) \wedge F_2(q_{n2})) \mid q_{i1} \in Q_1, q_{i2} \in Q_2, i = 0, 1, \ldots, n\} = (\bigvee\{I_1(q_{01}) \wedge \delta_1(q_{01}, a_1, q_{11}) \wedge \delta_1(q_{11}, a_2, q_{21}) \wedge \cdots \wedge \delta_1(q_{n-1,1}, a_n, q_{n1}) \wedge F_1(q_{n1}) \mid q_{i1} \in Q_1, i = 0, 1, \ldots, n\}, \bigvee\{I_2(q_{02}) \wedge \delta_2(q_{02}, a_1, q_{12}) \wedge \delta_2(q_{12}, a_2, q_{22}) \wedge \cdots \wedge \delta_2(q_{n-1,2}, a_n, q_{n2}) \wedge F_2(q_{n2}) \mid q_{i2} \in Q_2, i = 0, 1, \ldots, n\}) = (f_1(\omega), f_2(\omega))$.

This completes the proof that the given mapping $g : \Sigma^* \to L_1 \times L_2$ is a quantum reversible regular language.

Example 2. Let $(L, \leq, \vee, \wedge, \perp, 0, 1)$ be an orthormodular lattice, where $L = \{0, a, a^\perp, 1\}$. Quantum reversible fuzzy automata $\mathcal{A} = (Q_1, \Sigma, \delta_1, I_1, F_1)$ and $\mathcal{B} = (Q_2, \Sigma, \delta_2, I_2, F_2)$ over L are given as follows: $Q_1 = \{p_1, q_1\}$, $Q_2 = \{p_2, q_2\}$, $\Sigma = \{\sigma\}$, $I_1(p_1) = a$, $I_1(q_1) = 1$, $F_1(p_1) = 1$, $F_1(q_1) = a$, $I_2(p_2) = 1$, $I_2(q_2) = 0$, $F_2(p_2) = a^\perp$, $F_2(q_2) = a$, a quantum transition relation $\delta_1 : Q_1 \times \Sigma \times Q_1 \to L$ is given by $\delta_1(p_1, \sigma, p_1) = 1$, $\delta_1(p_1, \sigma, q_1) = a$, $\delta_1(q_1, \sigma, p_1) = a^\perp$, $\delta_1(q_1, \sigma, q_1) = 0$, and another quantum transition relation $\delta_2 : Q_2 \times \Sigma \times Q_2 \to L$ is given by $\delta_2(p_2, \sigma, p_2) = a^\perp$, $\delta_2(p_2, \sigma, q_2) = a$, $\delta_2(q_2, \sigma, p_2) = a$ and $\delta_2(q_2, \sigma, q_2) = 0$. Their graphical representations are shown in Fig. 2. By computation, $\mathcal{A}$ and $\mathcal{B}$'s quantum reversible regular languages, denoted by $f_{\mathcal{A}}$ and $f_{\mathcal{B}}$ respectively, are $f_{\mathcal{A}}(\varepsilon) = a$, $f_{\mathcal{A}}(\omega) = 1$ for any $\omega \in \Sigma^+$; $f_{\mathcal{B}}(\varepsilon) = a^\perp$, $f_{\mathcal{B}}(\sigma) = 1$ and $f_{\mathcal{B}}(\omega) = a^\perp$ for any $\omega \in \Sigma^* \setminus \{\varepsilon, \sigma\}$.

By Proposition 4, there exists a quantum reversible fuzzy automaton $\mathcal{M} = (Q, \Sigma, \delta, I, F)$ over $L \times L$ such that its languages $f_{\mathcal{M}} = (f_{\mathcal{A}}, f_{\mathcal{B}})$, where $Q = \{p_{11}, p_{12}, p_{21}, p_{22}\}$, $p_{11} = (p_1, p_2)$, $p_{12} = (p_1, q_2)$, $p_{21} = (q_1, p_2)$, $p_{22} = (q_1, q_2)$, the quantum initial state $I : Q \to L \times L$ is given by $I(p_{11}) = (a, 1)$, $I(p_{12}) = (a, 0)$, $I(p_{21}) = (1, 1)$ and $I(p_{22}) = (1, 0)$; the quantum final state $F : Q \to L \times L$ is given by $F(p_{11}) = (1, a^\perp)$, $F(p_{12}) = (1, a)$, $F(p_{21}) = (a, a^\perp)$ and $F(p_{22}) = (a, a)$; and $\delta : Q \times \Sigma \times Q \to L \times L$ is given by $\delta(p_{11}, \sigma, p_{11}) = (1, a^\perp)$, $\delta(p_{12}, \sigma, p_{12}) = (1, 0)$, $\delta(p_{21}, \sigma, p_{21}) = (0, a^\perp)$, $\delta(p_{22}, \sigma, p_{22}) = (0, 0)$, $\delta(p_{11}, \sigma, p_{12}) = \delta(p_{12}, \sigma, p_{11}) = (1, a)$, $\delta(p_{21}, \sigma, p_{22}) = \delta(p_{22}, \sigma, p_{21}) = (0, a)$, $\delta(p_{11}, \sigma, p_{21}) = (a, a^\perp)$, $\delta(p_{21}, \sigma, p_{11}) = (a^\perp, a^\perp)$, $\delta(p_{22}, \sigma, p_{12}) = (a^\perp, 0)$ and $\delta(p_{12}, \sigma, p_{22}) = (a, 0)$. The graphical representation of $LQRFA$ $\mathcal{M}$ is shown in

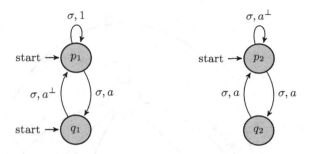

Fig. 2. Graphical representation of $LQRFAs$ $\mathcal{A}$ and $\mathcal{B}$ (from left to right)

Fig. 3. Its language is as follows: $f_{\mathcal{M}}(\varepsilon) = (a, a^{\perp})$, $f_{\mathcal{M}}(\sigma) = (1,1)$ and $f_{\mathcal{M}}(\omega) = (1, a^{\perp})$, for any $\omega \in \Sigma^{+} \setminus \{\sigma\}$.

Proposition 5. *Let mappings $f_1, f_2 : \Sigma^* \to L$ be two quantum reversible regular languages. Then there exist a quantum reversible automaton $\mathcal{M} = (Q, \Sigma, \delta, I, F)$ over $L \times L$ with the language $f_{\mathcal{M}}$ accepted and a mapping $h : L \times L \to L$ such that the composition of h and $f_{\mathcal{M}}$ is the intersection between f_1 and f_2, i.e., $h \circ f_{\mathcal{M}} = f_1 \cap f_2$.*

Proof. Let mappings $f_1, f_2 : \Sigma^* \to L$ be two quantum reversible regular languages. Then by Proposition 4, there exists a quantum reversible automaton $\mathcal{M} = (Q, \Sigma, \delta, I, F)$ over $L \times L$ such that whose accepted language is $f_{\mathcal{M}}(\omega) = (f_1(\omega), f_2(\omega)), \forall \omega \in \Sigma^*$. Construct a mapping $h : L \times L \to L$ as $h(x, y) = x \wedge y, \forall x, y \in L$. So $h \circ f_{\mathcal{M}}(\omega) = h(f_{\mathcal{M}}(\omega)) = h(f_1(\omega), f_2(\omega)) = f_1(\omega) \wedge f_2(\omega) = (f_1 \cap f_2)(\omega), \forall \omega \in \Sigma^*$. Hence $h \circ f_{\mathcal{M}} = f_1 \cap f_2$.

Example 3. As shown in Example 2, construct a mapping $h : L \times L \to L$ such that $h(u, v) = u \wedge v$ for any $u, v \in L$. Then $h \circ f_{\mathcal{M}}(\varepsilon) = a \wedge a^{\perp} = 0 = (f_{\mathcal{A}} \cap f_{\mathcal{B}})(\varepsilon)$, $h \circ f_{\mathcal{M}}(\sigma) = 1 = (f_{\mathcal{A}} \cap f_{\mathcal{B}})(\sigma)$, and $h \circ f_{\mathcal{M}}(\omega) = 1 \wedge a^{\perp} = a^{\perp} = (f_{\mathcal{A}} \cap f_{\mathcal{B}})(\omega)$ for all $\omega \in \Sigma^{+} \setminus \{\sigma\}$. Therefore, the composition of h and $f_{\mathcal{M}}$ is the intersection between $f_{\mathcal{A}}$ and $f_{\mathcal{B}}$.

Proposition 6. *Let $f \in l_{QRRGs}(\Sigma^*)$. If there is a quantum reversible fuzzy automaton $\mathcal{A} = (Q, \Sigma, \delta, I, F)$ over an orthomodular lattice L such that $\mathcal{A}$ accepts f, then there is a quantum reversible fuzzy automaton $\mathcal{B}$ such that $f_{\mathcal{B}}(\omega) = f(\omega^{-1})$, $\forall \omega \in \Sigma^*$.*

Proof. Construct a quantum fuzzy automaton $\mathcal{B} = (Q, \Sigma, \delta_B, I_B, F_B)$ over an orthomodular lattice L as follows: I_B is a mapping from Q to L, which is given by, $I_B(q) = F(q), \forall q \in Q$; a mapping $F_B(q) : Q \to L$ is defined as $F_B(q) = I(q), \forall q \in Q$; and a mapping $\delta_B : Q \times \Sigma \times Q \to L$ is given by $\delta_B(q, \sigma, p) = \delta(p, \sigma, q), \forall q, p \in Q, \sigma \in \Sigma$.

Claim 1. The quantum fuzzy automaton $\mathcal{B} = (Q, \Sigma, \delta_B, I_B, F_B)$ is reversible, which holds clearly by the quantum reversible fuzzy automaton $\mathcal{A} = (Q, \Sigma, \delta, I, F)$ over L.

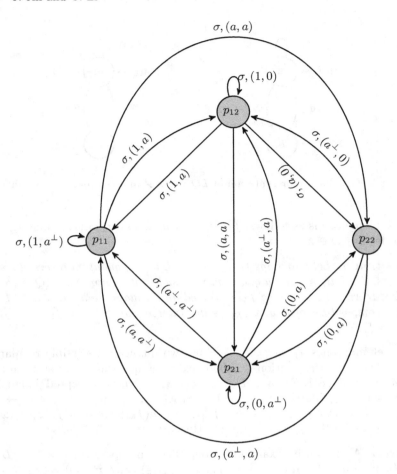

Fig. 3. Graphical representation of an $LQRFA$ $\mathcal{M}$

Claim 2. $f_{\mathcal{B}}(\omega) = f(\omega^{-1})$, $\forall \omega \in \Sigma^*$.

In fact, $f_{\mathcal{B}}(\varepsilon) = \bigvee\{I_B(q) \wedge F_B(q) \mid q \in Q\} = \bigvee\{I(q) \wedge F(q) \mid q \in Q\} = f(\varepsilon)$.

For any $\omega \in \Sigma^+$ with length $|\omega| = n$, let $\omega = a_1 \cdots a_n$, $a_i \in \Sigma, i = 1, \cdots, n$.

Then

$$f_{\mathcal{B}}(\omega) = \bigvee\{I_B(q_0) \wedge \delta_B(q_0, a_1, q_1) \wedge \delta_B(q_1, a_2, q_2) \wedge \cdots \wedge \delta_B(q_{n-1}, a_n, q_n) \wedge F_B(q_n) | q_0, q_1, \cdots, q_n \in Q\} = \bigvee\{F(q_0) \wedge \delta(q_1, a_1, q_0) \wedge \delta(q_2, a_2, q_1) \wedge \cdots \wedge \delta(q_n, a_n, q_{n-1}) \wedge I(q_n) | q_0, q_1, \cdots, q_n \in Q\} = \bigvee\{I(q_n) \wedge \delta(q_n, a_n, q_{n-1}) \wedge \cdots \wedge \delta(q_2, a_2, q_1) \wedge \delta(q_1, a_1, q_0) \wedge F(q_0) | q_0, q_1, \cdots, q_n \in Q\} = f_{\mathcal{A}}(\omega^{-1}) = f(\omega^{-1}).$$

This completes the proof from the above claims.

Example 4. Let $(L, \leq, \vee, \wedge, \perp, 0, 1)$ be an orthomodular lattice, where $L = \{0, a, a^{\perp}, 1\}$. An $LQRFA$ $\mathcal{A} = (Q, \Sigma, \delta_1, I_1, F_1)$ over L is given as follows: $Q = \{p_1, q_1\}$, $\Sigma = \{0, 1\}$, $I_1(p_1) = 1$, $I_1(q_1) = a$, $F_1(p_1) = 1$, $F_1(q_1) = 1$, a quantum transition relation $\delta_1 : Q \times \Sigma \times Q \to L$ is given by $\delta_1(p_1, 1, p_1) = a$, $\delta_1(p_1, 0, q_1) = a$, $\delta_1(q_1, 1, p_1) = a^\perp$, $\delta_1(q_1, 0, q_1) = a^\perp$, otherwise $\delta_1(q, \sigma, p) = 0$. By computation, $\mathcal{A}$'s quantum reversible regular languages, denoted by $f_{\mathcal{A}}$, is as follows: $f_{\mathcal{A}}(\varepsilon) = 1$, $f_{\mathcal{A}}(0) = f_{\mathcal{A}}(1) = f_{\mathcal{A}}(1^n) = f_{\mathcal{A}}(1^n 0) = a$ for any $n \geq 1$, otherwise, $f_{\mathcal{A}}(\omega) = 0$.

By Proposition 6, there exists an $LQRFA$ $\mathcal{B} = (Q, \Sigma, \delta_2, I_2, F_2)$ over L such that $f_{\mathcal{B}}(\omega) = f_{\mathcal{A}}(\omega^{-1})$, $\forall \omega \in \Sigma^*$, where $I_2(p_1) = I_2(q_1) = 1$, $F_2(p_1) = 1$, $F_2(q_1) = a$, and a quantum transition relation $\delta_2 : Q \times \Sigma \times Q \to L$ is given by $\delta_2(p_1, 1, p_1) = a$, $\delta_2(p_1, 1, q_1) = a^\perp$, $\delta_2(q_1, 0, q_1) = a^\perp$, $\delta_2(q_1, 0, p_1) = a$, otherwise $\delta_2(q, \sigma, p) = 0$. The graphical representations of $LQRFAs$ $\mathcal{A}$ and $\mathcal{B}$ are shown in Fig. 4. Clearly, $f_{\mathcal{B}}(\varepsilon) = 1$ and $f_{\mathcal{B}}(0) = f_{\mathcal{B}}(1) = f_{\mathcal{B}}(1^n) = f_{\mathcal{B}}(01^n) = a$ for any $n \geq 1$, otherwise, $f_{\mathcal{B}}(\omega) = 0$.

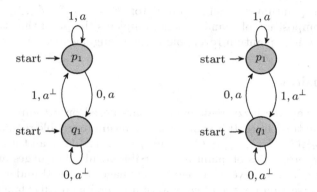

Fig. 4. Graphical representation of $LQRFAs$ $\mathcal{A}$ and $\mathcal{B}$ (from left to right)

Proposition 7. *Let $f \in l_{QRRGs}(\Sigma_2^*)$. If there is a homomorphism from Σ_1^* to Σ_2^*, i.e., $h : \Sigma_1^* \to \Sigma_2^*$, such that the restriction h on Σ_1 satisfies $h|_{\Sigma_1} \subseteq \Sigma_2$ and $h(\varepsilon) = \varepsilon$, then the composition of f and h is a quantum reversible regular language over Σ_1^*.*

Proof. Since f is a quantum reversible regular language over Σ_2^*, there exists a quantum reversible fuzzy automaton $\mathcal{A} = (Q, \Sigma_2, \delta, I, F)$ over an orthomodular lattice L such that $\mathcal{A}$ accepts f. Now construct a quantum fuzzy automaton $\mathcal{B} = (Q, \Sigma_1, \delta_B, I_B, F_B)$ over L as follows: $I_B = I$, $F_B = F$ and a mapping $\delta_B : Q \times \Sigma_1 \times Q \to L$ is given by $\delta_B(q, a, p) = \delta(q, h(a), p), \forall q, p \in Q, a \in \Sigma_1$. Obviously, the mapping δ_B is well defined since $h|_{\Sigma_1} \subseteq \Sigma_2$. Moreover, if $\delta_B(q, a, p) = \delta_B(q, a, p_1) > 0$ for any $q, p \in Q, a \in \Sigma_1$, then $\delta(q, h(a), p) = \delta(q, h(a), p_1)$ and so $p = p_1$ by the definition of the quantum reversible fuzzy automation $\mathcal{A}$. Similarly, if $\delta_B(q, a, p) = \delta_B(q_1, a, p) > 0$ for any $q, p \in Q, a \in \Sigma_1$, then $\delta(q, h(a), p) = \delta(q_1, h(a), p)$ and so $q = q_1$. Hence $\mathcal{B} = (Q, \Sigma_1, \delta_B, I_B, F_B)$ over L is a quantum reversible fuzzy automaton from Definition 2.

Noting that the composition of f and h is given by a mapping, $f \circ h : \Sigma_1^* \to L$, which is defined as

$$f \circ h(\omega) = f(h(\omega)), \forall \omega \in \Sigma_1^*.$$

Claim 1. $f_B(\varepsilon) = f \circ h(\varepsilon)$.
$f_B(\varepsilon) = \bigvee\{I_B(q) \wedge \delta_B^*(q, \varepsilon, p) \wedge F_B(p) \mid q, p \in Q\} = \bigvee\{I_B(q) \wedge F_B(q) \mid q \in Q\} = \bigvee\{I(q) \wedge F(q) \mid q \in Q\} = f(\varepsilon) = f(h(\varepsilon))$.
Claim 2. $f_B(\omega) = f \circ h(\omega), \forall \omega \in \Sigma_1^+$.
For any $\omega \in \Sigma^+$ with length $|\omega| = n$, let $\omega = a_1 \cdots a_n$, $a_i \in \Sigma, i = 1, \cdots, n$. Then

$f_B(\omega) = \bigvee\{I_B(q_0) \wedge \delta_B(q_0, a_1, q_1) \wedge \delta_B(q_1, a_2, q_2) \wedge \cdots \wedge \delta_B(q_{n-1}, a_n, q_n) \wedge F_B(q_n) | q_0, q_1, \cdots, q_n \in Q\} = \bigvee\{I(q_0) \wedge \delta(q_0, h(a_1), q_1) \wedge \delta(q_1, h(a_2), q_2) \wedge \cdots \wedge \delta(q_{n-1}, h(a_n), q_n) \wedge F(q_n) | q_0, q_1, \cdots, q_n \in Q\} = f(h(a_1)h(a_2) \cdots h(a_n))$
$= f(h(a_1 a_2 \cdots a_n)) = f \circ h(\omega)$.

Hence the quantum reversible automaton $\mathcal{B} = (Q, \Sigma_1, \delta_B, I_B, F_B)$ over L accepts the composition of f and h. This completes the proof that the composition of f and h is a quantum reversible regular language over Σ_1^*.

4   Conclusion

Quantum reversible fuzzy regular grammars based on orthomodular lattices are discussed in this paper. It is shown that quantum reversible regular languages among $l_{QRFRGs}(\Sigma^*)$, $l_{QRFHRGs}(\Sigma^*)$ and $l_{LQRFAs}(\Sigma^*)$ are all the same. Moreover, algebraic properties of quantum reversible regular languages are explored. The family of quantum reversible regular languages is closed under union operation. The intersection between two quantum reversible regular languages could be achieved by composition of a given mapping and some quantum reversible regular language. The reverse of quantum reversible regular language is also a quantum reversible regular language. The future work will further study Kleene closure of quantum reversible regular languages and their actual application.

References

1. Asveld, P.R.J.: Fuzzy context-free languages-part 1: generalized fuzzy context-free grammars. Theoret. Comput. Sci. **347**(1–2), 167–190 (2005)
2. Axelsen, H.B.: Clean translation of an imperative reversible programming language. In: Knoop, J. (ed.) CC 2011. LNCS, vol. 6601, pp. 144–163. Springer, Heidelberg (2011). https://doi.org/10.1007/978-3-642-19861-8_9
3. Axelsen, H.B., Glück, R.: A simple and efficient universal reversible turing machine. In: Dediu, A.-H., Inenaga, S., Martín-Vide, C. (eds.) LATA 2011. LNCS, vol. 6638, pp. 117–128. Springer, Heidelberg (2011). https://doi.org/10.1007/978-3-642-21254-3_8
4. Bennett, C.H.: Logical reversibility of computation. IBM J. Res. Dev. **17**(6), 525–532 (1973)
5. Chaudhari, S.R., Komejwar, D.D.: On fuzzy regular grammars. Adv. Fuzzy Math. **6**(1), 89–104 (2011)

6. Cheng, W., Wang, J.: Grammar theory based on quantum logic. Int. J. Theor. Phys. **42**(8), 1677–1691 (2003)
7. Costa, V.S., Bedregal, B.: Fuzzy linear automata and some equivalences. TEMA (São Carlos) **19**(1), 127–145 (2018)
8. Deufemia, V., Paolino, L., de Lumley, H.: Petroglyph recognition using self-organizing maps and fuzzy visual language parsing. In: 2012 IEEE 24th International Conference on Tools with Artificial Intelligence, vol. 1, pp. 852–859. IEEE (2012)
9. Li, L., Li, Y.: On reversible fuzzy automata. In: Quantitative Logic And Soft Computing, pp. 315–322. World Scientific (2012)
10. Li, Y.M.: Finite automata based on quantum logic and monadic second-order quantum logic. Sci. China Ser. F: Inf. Sci. **53**(1), 101–114 (2010)
11. Lombardy, S.: On the construction of reversible automata for reversible languages. In: Widmayer, P., Eidenbenz, S., Triguero, F., Morales, R., Conejo, R., Hennessy, M. (eds.) ICALP 2002. LNCS, vol. 2380, pp. 170–182. Springer, Heidelberg (2002). https://doi.org/10.1007/3-540-45465-9_16
12. MacLean, S., Labahn, G.: A new approach for recognizing handwritten mathematics using relational grammars and fuzzy sets. Int. J. Doc. Anal. Recogn. (IJDAR) **16**(2), 139–163 (2013)
13. Mordeson, J.N., Malik, D.S.: Fuzzy Automata and Languages: Theory and Applications. Chapman and Hall/CRC (2002)
14. Nielson, M.A., Chuang, I.L.: Quantum Computation and Quantum Information. Cambridge University Press, Cambridge (2000)
15. Feynman, R.P.: Quantum mechanical computers. Found. Phys. **16**(6), 507–531 (1986)
16. Shor, P.W.: Algorithms for quantum computation: discrete logarithms and factoring. In: Proceedings 35th Annual Symposium on Foundations of Computer Science, pp. 124–134. IEEE (1994)
17. Silva Farias, A.D., de Araújo Lopes, L.R., Bedregal, B., Santiago, R.H.N.: Closure properties for fuzzy recursively enumerable languages and fuzzy recursive languages. J. Intell. Fuzzy Syst. **31**(3), 1795–1806 (2016)
18. Thomsen, M.K., Glück, R., Axelsen, H.B.: Reversible arithmetic logic unit for quantum arithmetic. J. Phys. A: Math. Theor. **43**(38), 382002 (2010)
19. Van de Snepscheut, J.L.A.: What Is Computing All About? pp. 1–9. Springer, New York (1993).https://doi.org/10.1007/978-1-4612-2710-6
20. Xin, T., et al.: Nuclear magnetic resonance for quantum computing: techniques and recent achievements. Chin. Phys. B **27**(2), 020308 (2018)
21. Ying, M.: Automata theory based on quantum logic II. Int. J. Theor. Phys. **39**(11), 2545–2557 (2000)
22. Ying, M.: A theory of computation based on quantum logic (I). Theor. Comput. Sci. **344**(2–3), 134–207 (2005)
23. Yokoyama, T., Axelsen, H.B., Glück, R.: Principles of a reversible programming language. In: Proceedings of the 5th Conference on Computing Frontiers, pp. 43–54. ACM (2008)

A Signcryption Scheme Based Learning with Errors over Rings Without Trapdoor

Zhen Liu, Yi-Liang Han$^{(\boxtimes)}$, and Xiao-Yuan Yang

The Engineering University of People's Armed Police, Xi'an 710086, China
hanyil@163.com

Abstract. Signcryption has been extensively studied based on bilinear pairings, but rarely on lattices. This paper constructed a new lattice-based signcryption scheme in the random oracle model by skillfully combining a learning with errors over ring (R-LWE) based signature scheme (using the Fiat-Shamir with aborts technique) and a R-LWE based key exchange scheme, and then given a tight security reduction of strong unforgeability against adaptive chosen messages attacks (SUF-CMA) and indistinguishability against adaptive chosen ciphertext attacks (IND-CCA2) from the R-LWE problem to proposed scheme. This construction removed trapdoor by using the Fiat-Shamir with aborts technique, thus has a high efficiency. It needn't decrypt random coins for unsigncryption which may be used to construct multi-receiver signcryption.

Keywords: Lattice-based signcryption · Signcryption without trapdoor · Learning with errors over rings · Quantum attack resistant

1 Introduction

In the field of communication, cryptography is widely used for confidentiality and authentication. Signcryption that was first introduced by Zheng [1] is a important and elemental cryptographic primitive. It combines encryption and signature into one step, which can realize the confidentiality and authentication of information at the same time. Nowadays, there are a lot of research results on signcryption [2, 3]. However, most of them are based on discrete logarithm and factorization problems.

With the development of quantum computing, the security of traditional cryptographic schemes based on number theory is facing challenges. Cryptographic schemes against quantum attacks have attracted wide attention in academic.

The lattice has the advantages of high computational efficiency and powerful construction function (most cryptographic functions can be constructed based on lattices). Lattice-based cryptography [4] can resist quantum attacks, and has become the most potential representative of quantum attack-resistant cryptography. The research results of lattice-based encryption and signature are very fruitful. However, the lattice-based signcryption research is far from sufficient.

WHW12 [5] first constructs a lattice-based hybrid signcryption. Lattice-based signcryption [6–8] schemes were subsequently proposed. These schemes are based on the trapdoor generation algorithm and the preimage sample algorithm in the lattice, and the computational complexity of the algorithm is large, and also random coins need to

© Springer Nature Singapore Pte Ltd. 2019
X. Sun et al. (Eds.): NCTCS 2019, CCIS 1069, pp. 168–180, 2019.
https://doi.org/10.1007/978-981-15-0105-0_11

be decrypted in these processes of unsigncryption. This means that the random coins cannot be reused (Random coin reuse is the basis of constructing efficient multi-receiver signcryption) [9]. By using the Fiat-Shamir with aborts technique on lattices, LWWD16 [10] first proposed two lattice-based signcryption schemes without trapdoor generation algorithm and preimage sample algorithm, which both satisfy existence unforgeability against adaptive chosen messages attacks (EUF-CMA) authentication security, and achieved IND-CPA (indistinguishability against adaptive chosen plaintext attacks) and IND-CCA2 confidentiality security respectively in the random oracle model. It is a pity that the random coins also need to be decrypted in their IND-CCA2 scheme. Based on a ring version of the signature scheme [11] (also using the Fiat-Shamir with aborts technique on lattices), GM18 [12] proposed another lattice-based signcryption scheme in the random oracle model. However, it was still IND-CPA secure, and given no formal security proof of Unforgeability. Inspired by [13, 14], removing the use of trapdoor, WZGZ19 [15] construct an IND-CCA2 and strong unforgeability against adaptive chosen messages attacks (SUF-CMA) security lattice-based signcryption scheme, but it is identity-based construction.

Our Contribution. In this paper, we construct a new lattice-based signcryption scheme in the random oracle model by combining learning with errors over ring (R-LWE) based signature scheme in ABB16 [11] (using the Fiat-Shamir with aborts technique) and R-LWE based key exchange scheme in ADP16 [16], and then given a tight security reduction of strong unforgeability against adaptive chosen messages attacks (SUF-CMA) and IND-CCA2 from the R-LWE problem to our scheme. Our construction is based on signature-then-encryption mode proposed by Zheng [1]. Our scheme is the first lattice-based signcryption scheme that simultaneously satisfies SUF-CMA, IND-CCA2 security and needn't decrypt random coin for unsigncryption which may be used to construct multi-receiver signcryption.

2 Preliminaries

2.1 Notation

We denote the set of real numbers by $\mathbb{R}$, integers by $\mathbb{Z}$ and natural numbers by $\mathbb{N}$. We define $n = 2^k \in \mathbb{N}$ throughout the article for $k \in \mathbb{N}$, and also define that $q \in \mathbb{N}$ is a prime with $q = 1 \pmod{2n}$. All logarithms are in base 2. The finite field $\mathbb{Z}/q\mathbb{Z}$ is denoted by $\mathbb{Z}_q$. We identify an element in $\mathbb{Z}_q$ with its representative in $[-\lceil q/2 \rceil, \lfloor q/2 \rfloor]$ by writing (mod q). We define the ring $\mathcal{R} = \mathbb{Z}[x]/x^n + 1$ whose units set denoted by $\mathcal{R}^\times$, and define $\mathcal{R}_q = \mathbb{Z}_q[x]/x^n + 1$, $\mathcal{R}_{q,B} = \left\{ \sum_{i=0}^{n-1} a_i x^i \in \mathcal{R}_q | i \in [0, n-1], \ a_i \in [-B, B] \right\}$ for $B \in [0, q/2]$, and $\mathbb{B}_{n,\omega} = \{v \in \{0,1\}^n | \ \| v \|^2 = \omega\}$. We use lower-case letters to denote polynomials and lower-case bold letters (e.g., x) to express column vectors (x^T express row vector). We use upper-case bold letters to indicate matrices. For a vector x, we also use $\|x\|$ to denote Euclidean norm of x, and $O(\cdot)$ to denote Complexity.

2.2 Rounding Operators

We used the rounding operators that is defined in [16] for our construction. We use $[c]_{2^d}$ to denote the unique representative of c modulo 2^d in the set $\left(-2^{d-1}, 2^{d-1}\right]$ for $d \in \mathbb{N}$ and $c \in \mathbb{Z}$. We define rounding operator as $\lfloor \cdot \rceil_d : \mathbb{Z} \to \mathbb{Z}, c \mapsto \left(c - [c]_{2^d}\right)/2^d$. We naturally extend these definitions to vectors and polynomials by applying $\lfloor \cdot \rceil_d$ and $[\cdot]_{2^d}$ to each component of the vector and to each coefficient of the polynomial, respectively. We use $\lfloor v_{d,q} \rceil$ to express the abbreviation of $\lfloor v(\bmod q) \rceil_d$.

2.3 Lattices

Let $n \geq k > 0$. A k-dimensional lattice Λ is a discrete additive subgroup of $\mathbb{R}^n$ containing all integer linear combinations of k linearly independent vectors $\{\boldsymbol{b}_1, \boldsymbol{b}_2, \cdots, \boldsymbol{b}_k\} = \boldsymbol{B}$, i.e. $\Lambda = \Lambda(\boldsymbol{B}) = \{\boldsymbol{Bx} | \boldsymbol{x} \in \mathbb{Z}^k\}$. Throughout this paper we are mostly concerned with q-ary lattices. $\Lambda \in \mathbb{Z}^n$ is called a q-ary lattice if $q\mathbb{Z} \subset \Lambda$ for some $q \in \mathbb{Z}$. Let $\boldsymbol{A} \leftarrow_\$ \mathbb{Z}_q^{m \times n}$. We define the q-ary lattices $\Lambda_q^\perp(A) = \{\boldsymbol{x} \in \mathbb{Z}^n | \boldsymbol{Ax} = 0(\bmod q)\}$ and $\Lambda_q(A) = \{\boldsymbol{x} \in \mathbb{Z}^n | \exists \boldsymbol{s} \in \mathbb{Z}^m s.t. \boldsymbol{x} = \boldsymbol{A}^T \boldsymbol{s}(\bmod q)\}$. Furthermore, for $\boldsymbol{u} \in \mathbb{Z}_q^m$ we define cosets $\Lambda_{u,q}^\perp(A) = \{\boldsymbol{x} \in \mathbb{Z}^n | \boldsymbol{Ax} = \boldsymbol{u}(\bmod q)\}$, i.e., $\Lambda_q^\perp(A) = \Lambda_{0,q}^\perp(A)$.

2.4 Learning with Errors over Rings

Definition 1 (R-LWE Distribution). For an $s \in \mathcal{R}$ and a distribution χ over $\mathcal{R}$, a sample from the ring-LWE distribution $A_{s,\chi}$ over $\mathcal{R}_q \times \mathcal{R}_q$ is generated by choosing $a \leftarrow \mathcal{R}_q$ uniformly at random, choosing $e \leftarrow \chi$, and outputting $(a, t = as + e)$.

Definition 2 (Decisional R-LWE). The decision version of the R-LWE problem, denoted decisional R-LWE$_{q,n,m,\chi}$ is to distinguish with non-negligible advantage between independent m times samples from $A_{s,\chi}$, where $s \leftarrow \chi$ is chosen once and for all, and the same number of uniformly random and independent samples from $\mathcal{R}_q \times \mathcal{R}_q$. We also write R-LWE$_{q,n,\chi}$ when $m = 1$.

The R-LWE assumption comes with a worst-case to average-case reduction to problems over ideal lattices [18]. Furthermore, it was shown in [19] that the learning with errors problem remains hard if one chooses the secret distribution to be the same as the error distribution. We write R-LWE$_{q,n,m,\sigma}$ if σ is the discrete Gaussian distribution with standard deviation.

2.5 Reconciliation Mechanism

In this paper, we borrow the notations [16] methods to achieve the key exchange, who defines the **HelpRec** function to compute the k-bit reconciliation information for the key exchange. We just give a brief introduction as follows, more detail can be seen in [16].

In [16], they show how to agree on a n bit key from either a polynomial of degree $2n$ or $4n$ by **HelpRec** and **Rec** function.

HelpRec(x) taking as input a ring element and outputting a reconciliation vector r.

Rec$(x'; r)$ taking as input a ring element and a reconciliation vector and outputting a symmetric key K.

If x and x' are close to each other (the distance between their coefficients is small), the output of **Rec**$(x; r)$ and **Rec**$(x'; r)$ are the same.

Lemma D provides a detailed analysis of the failure probability of the key agreement and shows that it is smaller than 2^{-60} (see detail in [16]).

3 Signcryption: Model and Security

Definition 3 (Signcryption). For public parameters prm generated by security parameter k, a signcryption scheme SC consists of a four tuple of polynomial-time algorithm SC $=$ (KeyGen$_R$, KeyGen$_S$, SC, USC) as follows:

- KeyGen$_R$(prm): KeyGen$_R$ is a randomised key-generation algorithm of receivers that on input a public parameter prm, outputs a receiver's public-key pk_R and a receiver's secret-key sk_R.
- KeyGen$_S$(prm): KeyGen$_S$ is a randomised key-generation algorithm of senders that on input a public parameter prm, outputs a sender's public-key pk_S and a sender's secret-key sk_S.
- $SC(m, pk_R, sk_S)$: SC is a randomised signcrypt algorithm that on input a receiver's public-key pk_R, a sender's secret-key sk_S and a message $m \in \mathcal{M}$ ($\mathcal{M}$ is the message space), outputs a ciphertext C.
- $USC(C, pk_S, sk_R)$: USC is a deterministic unsigncrypt algorithm that on input a sender's public-key pk_S, a receiver's secret key sk_R and a ciphertext C, outputs a message m or an invalid $\perp$.

It is required that for any $(pk_S, sk_S) \leftarrow$ **KeyGen$_S$**(prm) and $(pk_R, sk_R) \leftarrow$ **KeyGen$_R$**(prm), $m \leftarrow$ **UnSigncrypt**(C, pk_S, sk_R), where $C \leftarrow$ **Signcrypt**(m, pk_R, sk_S), holds.

A signcryption scheme should satisfy confidentiality and authenticity, which correspond to IND-CCA2 and SUF-CMA.

For the IND-CCA2 property, we consider the following game played between a challenger B and an adversary A.

Initial: B runs the key generation algorithm to generate a receiver's public/private key pair (pk_R^*, sk_R^*) sends pk_R^* to A and keeps sk_R^* secret.

Phase 1: A can perform a polynomially bounded number of unsigncryption queries in an adaptive manner. In an unsigncryption query, A submits a ciphertext C with a sender's public key/private key pair (pk_S, sk_S) to B. B runs the unsigncryption oracle and returns the message **UnSigncrypt**(C, pk_S, sk_R^*), if it is a valid ciphertext. Otherwise, B returns rejection symbol $\perp$.

Challenge: B decides when Phase 1 ends. A chooses two equal length plaintexts (m_0, m_1) and a sender's public/private key pair (pk_S^*, sk_S^*), and sends these to B. B takes a random bit b from $\{0, 1\}$ and runs the signcryption oracle, which returns ciphertext $C^* =$ **Signcrypt**(m_b, pk_R^*, sk_S^*) to A as a challenged ciphertext.

Phase 2: A can ask a polynomially bounded number of unsigncryption queries adaptively again as in Phase 1 with the restriction that it cannot make an unsigncryption query on the challenged ciphertext C^* with the same sender's public/private key pair (pk_S^*, sk_S^*).

Guess: A produces a bit b' and wins the game if $b' = b$.

The advantage of A is defined as $Adv_{SC,A}^{CCA2}(k) = |2 \Pr[b' = b] - 1|$, where $\Pr[b' = b]$ denotes the probability that $b' = b$.

A signcryption scheme is (ϵ, t, q_u)-IND-CCA2 secure if no probabilistic t-polynomial time adversary A has advantage at least ϵ after at most q_u unsigncryption queries in the IND-CCA2 game.

Notice that A knows the sender's private key sk_S^* in the preceding definition. This condition corresponds to the stringent requirement of insider security for confidentiality of signcryption [16]. On the other hand, it ensures the forward security of the scheme, that is, confidentiality is preserved in case the sender's private key becomes compromised.

For the SUF-CMA property, we consider the following game played between a challenger D and an adversary F:

Initial: D runs the key generation algorithm to generate a sender's public/private key pair (pk_S^*, sk_S^*). D sends pk_S^* to F and keeps sk_S^* secret.

Attack: F can perform a polynomially bounded number of signcryption queries in an adaptive manner. In a signcryption query, F submits a message m and a receiver's public key/private key pair (pk_R, sk_R) to D. D runs the signcryption oracle, which returns the ciphertext D Signcrypt $C = \boldsymbol{Signcrypt}(m, pk_R, sk_S^*)$. Then, D sends C to F.

Forgery: At the end of the game, F chooses a receiver's public/private key pair (pk_R^*, sk_R^*) and produces a new ciphertext C^* of a plaintext m^* for the sender's public key pk_S^* (i.e., C^* was not produced by the signcryption oracle). F wins the game if C^* is a valid ciphertext.

The advantage of F is defined as the probability that it wins.

A signcryption scheme is (ϵ, t, q_s)-SUF-CMA secure if no probabilistic t-polynomial time adversary F has advantage at least ϵ after at most q_s signcryption queries in the SUF-CMA game. Note that the adversary F knows the receiver's private key sk_R^* in the preceding definition. Again, this condition corresponds to the stringent requirement of insider security for signcryption [16].

4   Signcryption Scheme Based on RLWE

4.1   Ring-LWE Based Signcryption Scheme (SC-RLWE)

Our signcryption scheme is parameterized by the integers $n \in \mathbb{N} > 0$, ω, d, B, q, U, L, l and the security parameter λ with $n > \lambda$, by the Gaussian distribution D_σ with standard deviation σ, by the hash function $H : \{0,1\}^* \rightarrow \{0,1\}^k$, by a random oracles $G : \{0,1\}^* \rightarrow \{0,1\}^l$ and by a reversible encoding function $F : \{0,1\}^k \rightarrow \mathbb{B}_{n,\omega}$. The encoding function F takes the (binary) output of the hash function H and maps it to a vector of length n and weight ω, and we use F^{-1} to represent the inverse function of F.

For more information about the encoding function see [17]. Furthermore, let $a_1, a_2 \in \mathcal{R}_q^\times$ be two uniformly sampled polynomials which are publicly known as global constants. They can be shared among arbitrarily many signers. We use prm to express the above public parameter.

$(pk_S, sk_S) \leftarrow \textbf{KeyGen}_S(prm):$

1: $x_S, e_{S1}, e_{S2} \leftarrow D_\sigma^n$
2: If $\text{CheckE}(e_{S1}) = 0 \vee \text{CheckE}(e_{S2}) = 0$, restart
3: $t_{S1} = a_1 x_S + e_{S1} (\text{mod } q), \ t_{S2} = a_2 x_S + e_{S2} (\text{mod } q)$
4: $sk_S \leftarrow x_S, e_{S1}, e_{S2}, \ pk_S \leftarrow t_{S1}, t_{S2}$
5: Return (pk_S, sk_S)

$(pk_R, sk_R) \leftarrow \textbf{KeyGen}_R(prm):$

1: $x_R, e_{R1}, e_{R2} \leftarrow D_\sigma^n$
1: $x_R, e_{R1}, e_{R2} \leftarrow D_\sigma^n$
2: If $\text{CheckE}(e_{R1}) = 0 \vee \text{CheckE}(e_{R2}) = 0$, restart
3: $t_{R1} = a_1 x_R + e_{R1} (\text{mod } q), \ t_{R2} = a_2 x_R + e_{R2} (\text{mod } q)$
4: $sk_R \leftarrow x_R, e_{R1}, e_{R2}, \ pk_R \leftarrow t_{R1}, t_{R2}$
5: Return (pk_R, sk_R)

$C \leftarrow \textbf{Signcrypt}(m, pk_R, sk_S)$

1: $y \leftarrow_\$ \mathcal{R}_{q,[B]}, \ y', y'', y''' \leftarrow D_\sigma^n$
2: $v_1 \leftarrow a_1 y + y', \ v_2 \leftarrow a_2 y + y''$
3: $u_1 \leftarrow t_{R1} y + y''', \ u_2 \leftarrow \text{HelpRec}(u_1)$
4: $K \leftarrow G(v_1, v_2, u_2, \text{Rec}(u_1, u_2), pk_S, pk_R)$
5: $c' \leftarrow H\left(\lfloor a_1 v_1 \rceil_{d,q}, \lfloor a_2 v_2 \rceil_{d,q}, m, pk_S, pk_R\right)$
6: $c \leftarrow F(c')$
7: $z \leftarrow x_S c + y$
8: $z_1 \leftarrow a_1 z + y' = a_1 x_S c + v_1, \ z_2 \leftarrow a_2 z + y'' = a_2 x_S c + v_2$
9: $w_1 \leftarrow a_1 v_1 - a_1 e_{s1} c, \ w_2 \leftarrow a_2 v_2 - a_2 e_{s2} c$
10: if $[w_1]_{2^d} = \lfloor a_1 v_1 \rceil_{d,q}, [w_2]_{2^d} = \lfloor a_2 v_2 \rceil_{d,q}$ and $z \in \mathcal{R}_{q,[B-U]}$ are not satisfied, restart.
11: $\mathcal{E} \leftarrow K \oplus (m \parallel z_1 \parallel z_2 \parallel c')$
12: Return $C \leftarrow (v_1, v_2, u_2, \mathcal{E})$

$m \leftarrow \textbf{UnSigncrypt}(C, pk_S, sk_R)$

1: $K \leftarrow G(v_1, v_2, u_2, \text{Rec}(x_R v_1, u_2), pk_S, pk_R)$
2: $m \parallel z_1 \parallel z_2 \parallel c' \leftarrow K \oplus \mathcal{E}$
3: $c \leftarrow F(c')$
4: $w_1 \leftarrow a_1 z_1 - a_1 t_{s1} c, \ w_2 \leftarrow a_2 z_2 - a_2 t_{s2} c$
5: return m if $c' = H\left(\lfloor w_1 \rceil_{d,q}, \lfloor w_2 \rceil_{d,q}, m, pk_S, pk_R\right)$ and $z_1, z_2 \in \mathcal{R}_q$, else $\perp$.

Correctness. If $C = (v_1, u, \mathcal{E})$ is a valid signcryption text, it is easy to see that $\text{Rec}(x_R v_1, u_2) = \text{Rec}(u_1, u_2)$ by Reconciliation mechanism in Sect. 2 (see [16] for detail), thus $K \leftarrow G(v_1, v_2, u_2, \text{Rec}(u_1, u_2), pk_S, pk_R)$ is hold at the process of unsigncryption. Afterwards, we can get the message and its signature $m \parallel z_1 \parallel z_2 \parallel c'$ by computing $K \oplus \mathcal{E}$. Further on, we compute $c = F(c')$ and it holds that $a_i z_i - a_i t_{si} c = a_i^2 y + a_i y' - a_i e_{si} c = a_i v_i - a_i e_{si} c$. So the two polynomials $w_i = a_i z_i - a_i t_{si} c$ can be computed. Because it satisfies that $\lfloor w_i \rfloor_{d,q} = \lfloor a_i y \rfloor_{d,q}$ for $i = 1, 2$ by rejection sampling in **Signcrypt** (see [11] for detail), we can check whether $c' = H\left(\lfloor w_1 \rfloor_{d,q}, \lfloor w_2 \rfloor_{d,q}, m, pk_S, pk_R \right)$. If this is satisfied, the signature $(z_1 \parallel z_2 \parallel c')$ is valid and the message m is returned.

4.2 Security Reduction

Theorem 1. Let $n, d, q, \omega, \sigma, B, U, L$ and l be arbitrary parameters satisfying the constraints described in above scheme. Assume that the Gaussian heuristic holds for lattice instances defined by the parameters above. if an adversary A has non-negligible advantage ϵ against the IND-CCA2 security of SC-RLWE when running in polynomial time t_A and performing q_{SC} signcryption queries, q_{DSC} deigncryption queries, q_H queries to oracles H and q_G queries to oracles G, then there is an algorithm B that breaks the Decisional R-LWE$_{q,n,\chi}$ problem (in the random oracle model) with success probability $\epsilon' > \epsilon - q_{DSC}(q_H/2^n + q_G/2^l)$ and within the running time $t_B = t_A + \mathcal{O}(q_{SC}k^2 + q_H + q_G)$.

Proof: Suppose that a simulator is given a distribution (a, t) from either distribution $A_{s,\chi}$ as definition 2.5 where χ express a Gaussian distribution D_σ^n over $\mathcal{R}_q$ and $s \leftarrow \chi$ is chosen once and for all, or uniformly random and independent from $\mathcal{R}_q \times \mathcal{R}_q$. We show how to use this adversary A to construct a distinguisher algorithm B for the Decisional R-LWE$_{q,n,\chi}$ problem as follows.

Prepare the Public Key. For given parameters and a distribution (a, t) as above theorem, the simulator chooses $a' \in \mathcal{R}_q^\times$ uniformly at random, sets $a_1 = a$ and computes $a_2 = a'a$, then sets a_1, a_2 as public parameter, afterwards picks $e_{S2}' \leftarrow D_\sigma^n$, replaces $t_{R1} = t$, computes $t_{R2}^* = a't + e_{S2}'$. Finally the simulator Outputs public key $pk_R^* = \{ t_{R1}^* = t, t_{R2}^* \}$, and sends it to adversary A.

Hash Queries and Unsigncryption Query of the First Stage. H simulation for (m, pk_S, sk_S). When a hash query of H is received, the simulator checks if the query tuple $(m, pk_S, sk_S, pk_R^*, z_1, z_2, c')$ is already in L_1. If it exists, the result of c' is returned. Else, the simulator picks a random number $z \leftarrow \mathcal{R}_{q,[B-U]}, y', y'' \leftarrow D_\sigma^n$ and $c'' \in \{0, 1\}^k$, computes $c = F(c'')$, $z_1 = a_1 z + y'$, $z_2 = a_2 z + y''$, $w_1 = a_1 z_1 - a_1 t_{s1} c$ and $w_2 = a_2 z_2 - a_2 t_{s2} c$. Then it checks whether $\lfloor w_{1j} \rfloor_{2^d} < 2^d - L$ and $\lfloor w_{2j} \rfloor_{2^d} < 2^d - L$ for all $j \in (1, 2, \cdots n)$. If it is not satisfied, restart, else by rejection sampling, we have $\lfloor w_i \rfloor_{d,q} = \lfloor a_i y \rfloor_{d,q}$ for $i = 1, 2$ (see detail in [11]). So the simulator can compute $c' = H\left(\lfloor w_1 \rfloor_{d,q}, \lfloor w_2 \rfloor_{d,q}, m, pk_S, pk_R^* \right)$, then insert the tuple $(m, pk_S, sk_S, pk_R^*, z_1, z_2, c')$ into list L_1 and returns the result of c'.

G simulation for $(v_1, v_2, u_2, pk_S, sk_S)$. When a hash query of G is received, the simulator checks if the query tuple $(v_1, v_2, u_2, pk_S, sk_S, pk_R^*, K)$ is already in L_2. If it exists, the existing result of K is returned. If it does not exist, the simulator picks a random number $x \in D_\sigma^n$, compute $K = G(v_1, v_2, u_2, \text{Rec}(xv_1, u_2), pk_S, pk_R^*)$, inserts the tuple $(v_1, v_2, u_2, pk_S, sk_S, pk_R^*, K)$ into list L_2 and returns the number K.

Unsigncryption query simulation for $(v_1, v_2, u_2, \mathcal{E}, pk_S, sk_S)$. When Unsigncryption query of $(v_1, v_2, u_2, \mathcal{E}, pk_S, sk_S)$ is received, the simulator traverses L_1 and L_2 to check $(m, pk_S, sk_S, pk_R^*, z_1, z_2, c')$ and $(v_1, v_2, u_2, pk_S, sk_S, pk_R^*, K)$, such that $\mathcal{E} \leftarrow K \oplus (m \parallel z_1 \parallel z_2 \parallel c')$. If such a tuple exists, return m, otherwise return $\perp$.

Prepare the Challenge Ciphertext. After completing the first stage, A chooses two messages m_0 and m_1 in $\mathcal{M}$ together with an arbitrary sender's key (pk_S^*, sk_S^*), asks a challenge signcryption text produced by the simulator under receiver's public key $pk_R^* = \{t_{R1}^* = t, t_{R2}^*\}$, then the simulator picks $b \in \{0, 1\}$ randomly, performs as follows.

It chooses $y \leftarrow_\$ \mathcal{R}_{q,[B]}$, $y', y'', y''' \leftarrow D_\sigma^n$, then computes $v_1^* = ay + y'$, $v_2^* = a'ay + y''$, $u_2^* = ty + y'''$, $u_2^* = \text{HelpRec}(u_1^*)$, $K^* = G(v_1^*, v_2^*, u_2^*, \text{Rec}(u_1^*, u_2^*), pk_S^*, pk_R^*)$, $c'^* = H(\lfloor v_1^* \rceil_{d,q}, \lfloor v_2^* \rceil_{d,q}, m_b, pk_S^*, pk_R^*)$, next encodes $c^* = F(c'^*)$, and then computes $z_1^* = ax_S^*c + v_1^*$, $z_2^* = a'ax_S^*c + v_2^*$, $w_1^* = az_1^* - ae_{s1}^*c$ and $w_2^* = a'az_2^* - a'ae_{s2}^*c$. If $[w_1^*]_{2^d}$, $[w_2^*]_{2^d} \notin \mathcal{R}_{q,[2^d - L]}$ and $z_1^*, z_2^* \notin \mathcal{R}_q$, restart, else computes $\mathcal{E}^* = K^* \oplus (m_b \parallel z_1^* \parallel z_2^* \parallel c'^*)$, finally, it outputs $C^* = (v_1^*, v_2^*, u_2^*, \mathcal{E}^*)$ and sends it to adversary A.

Hash Queries and Unsigncryption Query of the Second Stage. Adversary A repeats the first stage operations, but he can't ask unsigncryption on $C^* = (v_1^*, v_2^*, u_2^*, \mathcal{E}^*)$.

The Guess Stage. Adversary A outputs $b' \in \{0, 1\}$ as his guess of b.

As can be seen from the above, the simulation is perfect. When the input distribution (a, t) to the simulator comes from distribution $A_{s,\chi}$, the output of the encryption oracle is just a perfectly legitimate ciphertext. It is clear that the joint distribution of the adversary's view and b is identical to that in the actual attack. However, when the input distribution (a, t) to the simulator uniformly random samples from $\mathcal{R}_q \times \mathcal{R}_q$, the adversary's view and b are essentially information theoretic independent. This completes the construction of a distinguisher for the Decisional R-LWE$_{q,n,\chi}$ problem.

Now we analyze the advantages. The only event that causes the simulation not perfect is that a valid signcryption text be rejected in unsigncryption query stage. It is the result of simulation of H and G. For the queries on H, the probability is no more than $q_H/2^k$. For the queries on G, the probability is no more than $q_G/2^l$. Hence, the probability of algorithm B breaks the Decisional R-LWE$_{q,n,\chi}$ problem is $\epsilon \geq \epsilon - q_{USC}(q_H/2^k + q_G/2^l)$.

Then we consider the running time. As can be seen from the above, the operations of algorithm B are all efficient. We already showed that, the distributions of the signcryption simulated by B and those obtained by the actual signcrypting algorithm are statistically close. Thus, when emulating the signcryption procedure, B rejects a pair $(v_1, v_2, u_2, \mathcal{E})$ with the same probability as when running algorithm **UnSigncrypt** is executed. More precisely, simulating the signcrypting process essentially consists of a number of polynomial multiplications, on average $\mathcal{O}\left((k+l)^2\right)$ per query, which leads to the approximate bound $t_D = t_A + \mathcal{O}\left(q_{USC}(k+l)^2 + q_H + q_G\right)$.

Theorem 2. Let $n, d, q, \omega, \sigma, B, U, k, L$ and l be arbitrary parameters satisfying the constraints described in above scheme. Assume that the Gaussian heuristic holds for lattice instances defined by the parameters above. For every SUF-CMA adversary A that runs in time t_A, performing at most q_{SC} signcryption queries, q_H queries to oracles H and q_G queries to oracles G, respectively, and forges a valid signcryption of our SC-RLWE scheme with probability ϵ_A, there exists a distinguisher D that runs in time $t_D = t_A + \mathcal{O}\left(q_{SC}(k+l)^2 + q_H + q_G\right)$ and breaks the Decisional R-LWE$_{q,n,2,\sigma}$ problem (in the random oracle model) with success probability

$$\epsilon_D \geq \epsilon_A \left(1 - \frac{(q_H + q_G)q_{SC}2^{2n(d+1)}}{(2B+1)^n q^n}\right) - \frac{2^{2dn}(q_H + q_G)(2B - 2U + 1)^n + (28\sigma + 1)^{3n}}{q^{2n}}.$$

Proof: Let A be an algorithm that runs in time t_A, makes at most q_{SC} signcryption queries, q_H queries to oracles H and q_G queries to oracles G, and forges a valid signcryption with probability ϵ_A. We show how to build a distinguisher D, solving Decisional R-LWE$_{q,n,2,\sigma}$ problem in time t_D and with probability ϵ_D as in the theorem statement.

Prepare the Public Key. For given parameters and two R-LWE$_{q,n,2,\sigma}$ challenge tuples (a_1, t_1) and (a_2, t_2), D simulates the ufcma-experiment for A, sets a_1, a_2 as public parameter, and sends $pk_S^* = (t_{S1}^* = t_1, t_{S2}^* = t_2)$ to the forger A. Then, it answers hash and signcrypting queries as follows.

Hash Queries. Hash oracle H simulation for (m, pk_R). When a hash query of H is received, the simulator checks if the query tuple $(p_1, p_2, m, pk_S^*, pk_R, c')$ is already in L_1. If it exists, the result of c' is returned. Else, the simulator picks a random number $c' \in \{0,1\}^k$, inserts the tuple $(p_1, p_2, m, pk_S^*, pk_R, c')$ into list L_1 and returns the result of c'.

Hash oracle G is simulated as in the proof of Theorem 1.

SC Queries. For a signcryption query on (m, pk_R), D simulates a signcryption $(v_1, v_2, u_2, \mathcal{E})$ on m by implementing the following steps:

D chooses $y \leftarrow_\$ \mathcal{R}_{q,[B]}$ and $y', y'', y''' \leftarrow D_\sigma^n$, then computes $v_1 = a_1 y + y'$, $v_2 \leftarrow a_2 y + y''$, $u_1 \leftarrow t_{R1}y + y'''$, $u_2 = \text{HelpRec}(u_1)$ and $K \leftarrow G(v_1, v_2, u_2\text{Rec}(u_1, u_2),$ $pk_S^*, pk_R)$. Afterwards it picks a random number $c' \in \{0,1\}^k$, a polynomial $z \in \mathcal{R}_{q,[B-U]}$, and computes $c = F(c')$. Next, it computes polynomials $z_1 = a_1 z + y'$, $z_2 = a_2 z + y''$, $w_1 = a_1 z_1 - a_1 t_1 c$ and $w_2 = a_2 z_2 - a_2 t_2 c$, then checks whether $[w_{1j}]_{2^d} < 2^d - L$ and $[w_{2j}]_{2^d} < 2^d - L$ for all $j \in (1, 2, \cdots n)$. If it is not satisfied, restart. Moreover, if the hash oracle H was queried before on $(\lfloor w_1 \rceil_{d,q},$ $\lfloor w_2 \rceil_{d,q}, m, pk_S^*, pk_R)$, then D aborts the simulation. Otherwise, D computes

$$c' = H\left(\lfloor w_1 \rceil_{d,q}, \lfloor w_2 \rceil_{d,q}, pk_S^*, pk_R\right), \qquad \mathcal{E} = K \oplus (m \parallel z_1 \parallel z_2 \parallel c'), \qquad \text{and} \qquad \text{returns}$$
$(v_1, v_2, u_2, \mathcal{E})$.

Forgery Signcryption Stage. Eventually A outputs a pair of receipient's keys $\left(pk_R^*, Sk_R^*\right)$ and a forgery $C^* = \left(v_1^*, v_2^*, u_2^*, \mathcal{E}^*\right)$. It required that the attacker A not to ask the Unsigncryption Oracle for C^*.

If the algorithm $\boldsymbol{UnSigncrypt}\left(C^*, pk_S^*, Sk_R^*\right)$ returns m^*, D returns 1, else D returns 0.

By the same Method of proof as Theorem 1 in [11], we can show that the responses to hash and signcrypting queries provided by the simulation D are indistinguishable from the random oracle's and signcrypting oracle's responses. We follow give the detail description.

The reduction emulates a random oracle, hence its answers are truly random as long as some given hash value was not already set. However, in the latter case the reduction aborts. It remains to show that (1) the distributions of the simulated signcryptions and the ones produced by the signcrypting algorithm (RLWE-SC) are statistically close, and that (2) the simulation does not abort too often.

(1) We apply Lemma 1 in [11] to show that the distribution of $z \in \mathcal{R}_{q,[B-U]}$ computed by the signcrypting algorithm (RLWE-SC) is statistically close to the uniform distribution on $\mathcal{R}_{q,[B-U]}$. To apply Lemma 1 in [11] we take X to be the uniform distribution on $\mathcal{R}_{q,[B-U]}$ with $U = 14\sigma\sqrt{\omega}$ and $B = 14\sigma n\sqrt{\omega}$. Furthermore, Z_v is the uniform distribution on the polynomials with coefficients in $[-B, B]^n + v$, that means, $z \leftarrow Z_v$ can be written as $z = v + y$, where $y \leftarrow_\$ \mathcal{R}_{q,[B]}$. Each element v is a polynomial with coefficient representation $v = Sc$ with entries Gaussian distributed with standard deviation $\sigma\sqrt{\omega}$. Hence, the coefficients of v are bounded by $14\sigma\sqrt{\omega}$ with overwhelming probability. By definition of the density function f_X and $f_{Z_{sc}}$, Eq. 1 in Lemma 1 of [11] overwhelming probability for $1/M = 1/e \approx \left(\frac{2(B-U)+1}{2B+1}\right)^n$ Hence, the hypothesis of Lemma 1 in [11] are fulfilled.

(2) Assume that the signcrypting simulation samples an additional value y uniformly at random in $\mathcal{R}_{q,[B]}$, and chooses y', y'', y''' at random in D_σ^n over $\mathcal{R}_q$. Furthermore, the simulation programs not only $c' = H\left(\lfloor w_1 \rceil_{d,q}, \lfloor w_2 \rceil_{d,q}, pk_S^*, pk_R\right)$ but also $c' = H\left(\lfloor a_1 v_1 \rceil_{d,q}, \lfloor a_2 v_2 \rceil_{d,q}, pk_S^*, pk_R\right)$. Sampling y does not influence (z, c'). It is clear that the abort probability during the simulation with the changes just described is the upper bound probability of aborting during the original signcrypting simulation. The former probability is the same as finding a collision. Hence, by Lemma 2 in [11] and the definition of rounding operators, the upper bound probability that D will abort during the simulation of A's environment is $q_{SC}(q_H + q_G + q_{SC})\frac{2^{2n(d+1)}}{(2B+1)^n q^n}$.

We next to show how to lower bound of D's distinguishing advantage in the R-LWE game with A's forging advantage against our scheme. Assume that A asks the query $\left(\lfloor z_1^* - t_1 c^* \rceil_{d,q}, \lfloor z_2^* - t_2 c^* \rceil_{d,q}, m^*, pk_S^*, pk_R^*\right)$ to the hash oracle. In the following, we distinguish between two cases: both t_1 and t_2 follow the R-LWE distribution (i.e., $t_1 = a_1 x_S^* + e_{S1}^* (\mathrm{mod}\, q)$ and $t_2 = a_2 x_S^* + e_{S2}^* (\mathrm{mod}\, q)$) or they are sampled uniformly at random in $\mathcal{R}_q$.

1st Case: The possibility that D outputs 0 also have two cases. One is that D aborts when answering a signcrypting query, and another is that the algorithm A does not output a valid forgery. The latter occurs with probability $1 - \epsilon_A$. The probability that D aborts during the simulation is $(q_H + q_G) \frac{2^{2n(d+1)}}{(2B+1)^n q^n}$. Thus, D returns 1 with probability at least $\epsilon_A q_{SC} (q_H + q_G) \frac{2^{2n(d+1)}}{(2B+1)^n q^n}$.

2nd Case: By Remark 1 in [11], we can bind the coefficients of the polynomials $x_S^*, e_{S1}^*, e_{S2}^*$ by 14σ with overwhelming probability. By Lemma 3 in [11] the probability that there exist polynomials with coefficient representation $x_S^*, e_{S1}^*, e_{S2}^* \in [-14\sigma, 14\sigma]^n$ is smaller than or equal to $\frac{(28\sigma + 1)^{3n}}{q^{2n}}$, such that $t_1 = a_1 x_S^* + e_{S1}^* (\mathrm{mod}\, q)$ for $t_1 \leftarrow \mathcal{R}_q$ and $t_2 = a_2 x_S^* + e_{S2}^* (\mathrm{mod}\, q)$ for $t_2 \leftarrow \mathcal{R}_q$. We suppose t_1, t_2 are not of this form. By the assumption that a hash query on $\left(\lfloor a_1 z_1^* - a_1 t_1 c^* \rceil_{d,q}, \lfloor a_2 z_2^* - a_2 t_2 c^* \rceil_{d,q}, m^*, pk_S^*, pk_R^* \right)$ is made by A, where $c^* = F(c'^*)$ and (z_1^*, z_2^*, c'^*) get by computing $K^* \oplus \mathcal{E}^*$ is A's forgery on message m^*, it suffices to show that for every hash query $c' = H(p_1, p_2, pk_S^*, pk_R)$ for $p_1, p_2 \in \mathcal{R}_q$, posed by A we can bound the probability that there exists a polynomial z such that (z_1, z_2, c') is a valid signature on m. That means, it is enough to show that for $p_1, p_2 \in \mathcal{R}_q$. We can bound the probability that for $c' \leftarrow_\$ \{0,1\}^k$ and $c = F(c')$, there exists $z \in \mathcal{R}_q$ such that $\| z \|_\infty \leq B - U$ and $p_i = \lfloor a_i z_i - a_i t_{si} c \rceil_{d,q}$, for $i = 1, 2$. By Corollary 1 in [11] the probability that such a polynomial z exists is smaller than $\frac{2^{2dn}(2B - 2U + 1)^n}{q^{2n}}$. So the probability can be bound by $\frac{(q_H + q_G) 2^{2dn}(2B - 2U + 1)^n}{q^{2n}}$ for that A forges a valid signcryption. Hence, the upper bound probability that D returns 1 is $\frac{(q_H + q_G) 2^{2dn}(2B - 2U + 1)^n + (28\sigma + 1)^{3n}}{q^{2n}}$.

Finally, $Adv_{n,q,\sigma}^{R-LWE}(A) = \epsilon_D \geq \epsilon_A \left(1 - \frac{(q_H + q_G) q_{SC} 2^{2n(d+1)}}{(2B+1)^n q^n} \right) - \frac{2^{2dn}(q_H + q_G)(2B - 2U + 1)^n + (28\sigma + 1)^{3n}}{q^{2n}}$.

It is clear that A's runtime t_A and D's runtime t_D are close. As D executes A as a subprocedure, we have $t_D \geq t_A$. The overhead for D is due to the extra steps needed to emulate the unforgeability game for A. We already showed that, the distributions of the signcryption simulated by D and those obtained by the actual signcrypting algorithm are statistically close. Thus, when emulating the signcryption procedure, D rejects a pair $(v_1, v_2, u_2, \mathcal{E})$ with the same probability as when running algorithm **Signcrypt** is executed. More precisely, simulating the signcrypting process essentially consists of a number of polynomial multiplications, on average $\mathcal{O}(k^2)$ per query, which leads to the approximate bound $t_D = t_A + \mathcal{O}(q_{SC} k^2 + q_H + q_G)$.

5   Conclusion

Lattice-based signcryption research is far from abundant. Constructing lattice-based signcryption without trapdoor is of great significance to enrich the lattice cryptography and promote the development of signcryption scheme. In this paper, we mainly focused on the security against quantum attacks, and proposed a new lattice-based signcryption scheme without trapdoor in the random oracle model by skillfully combining signature scheme in ABB16 and R-LWE based key exchange scheme in ADP16. We also gave a tight security reduction of SUF-CMA and IND-CCA2 from the R-LWE problem to proposed scheme. It needn't decrypt random coins for unsigncryption which may be used to construct multi-receiver signcryption.

Acknowledgement. This work was supported by National Natural Science Foundation of China (Grant Nos. 61572521, U1636114, 61772550), National Key R&D Program of China (Grant No. 2017YFB0802000), National Cryptography Development Fund of China (Grant No. MMJJ20170112), State Key Laboratory of Information Security (2017-MS-18).

References

1. Zheng, Y.: Digital signcryption or how to achieve cost (signature & encryption) ≪ cost (signature) + cost(encryption). In: Kaliski, B.S. (ed.) Advances in Cryptology — CRYPTO 1997. LNCS, vol. 1294, pp. 165–179. Springer, Heidelberg (1997). https://doi.org/10.1007/BFb0052234
2. Zheng, Y., Imai, H.: How to construct efficient signcryption schemes on elliptic curves. Inf. Process. Lett. **68**(5), 227–233 (1998)
3. Steinfeld, R., Zheng, Y.: A signcryption scheme based on integer factorization. In: Goos, G., Hartmanis, J., van Leeuwen, J., Pieprzyk, J., Seberry, J., Okamoto, E. (eds.) ISW 2000. LNCS, vol. 1975, pp. 308–322. Springer, Heidelberg (2000). https://doi.org/10.1007/3-540-44456-4_23
4. Micciancio, D., Regev, O.: Lattice-based cryptography. In: Bernstein D.J., Buchmann J., Dahmen E. (eds.) Post-Quantum Cryptography, pp. 147–191. Springer, Heidelberg (2009). https://doi.org/10.1007/978-3-540-88702-7_5
5. Wang, F., Hu, Y., Wang, C.: Post-quantum secure hybrid signcryption from lattice assumption. Appl. Math. Inf. Sci. **6**(1), 23–28 (2012)
6. Li, F., Bin Muhaya, F.T., Khan, M.K., et al.: Lattice-based signcryption. Concur. Comput.: Pract. Exp. **25**(14), 2112–2122 (2013)
7. Yan, J., Wang, L., Wang, L., et al.: Efficient lattice-based signcryption in standard model. Math. Probl. Eng. **2013**, 18 (2013)
8. Lu, X., Wen, Q., Jin, Z.: A lattice-based signcryption scheme without random oracles. Front. Comput. Sci. **8**(4), 667–675 (2014)
9. Bellare, M., Boldyreva, A., Staddon, J.: Randomness re-use in multi-recipient encryption schemeas. In: Desmedt, Y.G. (ed.) PKC 2003. LNCS, vol. 2567, pp. 85–99. Springer, Heidelberg (2003). https://doi.org/10.1007/3-540-36288-6_7
10. Lu, X., Wen, Q., Wang, L.: A lattice-based signcryption scheme without trapdoors. J. Electron. Inf. **38**(9), 2287–2293 (2016)
11. Akleylek, S., Bindel, N., Buchmann, J., Krämer, J., Marson, G.A.: An efficient lattice-based signature scheme with provably secure instantiation. In: Pointcheval, D., Nitaj, A., Rachidi,

T. (eds.) AFRICACRYPT 2016. LNCS, vol. 9646, pp. 44–60. Springer, Cham (2016). https://doi.org/10.1007/978-3-319-31517-1_3

12. Gérard, F., Merckx, K.: SETLA: signature and encryption from lattices. In: Camenisch, J., Papadimitratos, P. (eds.) CANS 2018. LNCS, vol. 11124, pp. 299–320. Springer, Cham (2018). https://doi.org/10.1007/978-3-030-00434-7_15

13. Lyubashevsky, V.: Lattice signatures without trapdoors. In: Pointcheval, D., Johansson, T. (eds.) EUROCRYPT 2012. LNCS, vol. 7237, pp. 738–755. Springer, Heidelberg (2012). https://doi.org/10.1007/978-3-642-29011-4_43

14. Tian, M., Huang, L.: Identity-based signatures from lattices: simpler, faster, shorter. Fundamenta Informaticae **145**(2), 171–187 (2016)

15. Wang, X., Zhang, Y., Gupta, B.B., et al.: An identity-based signcryption on lattice without trapdoor. J. UCS **25**(3), 282–293 (2019)

16. Alkim, E., Ducas, L., Pöppelmann, T., et al.: Post-quantum key exchange—a new hope. In: 25th {USENIX} Security Symposium ({USENIX} Security 16), pp. 327–343 (2016)

17. Güneysu, T., Lyubashevsky, V., Pöppelmann, T.: Practical lattice-based cryptography: a signature scheme for embedded systems. In: Prouff, E., Schaumont, P. (eds.) CHES 2012. LNCS, vol. 7428, pp. 530–547. Springer, Heidelberg (2012). https://doi.org/10.1007/978-3-642-33027-8_31

18. Applebaum, B., Cash, D., Peikert, C., Sahai, A.: Fast cryptographic primitives and circular-secure encryption based on hard learning problems. In: Halevi, S. (ed.) CRYPTO 2009. LNCS, vol. 5677, pp. 595–618. Springer, Heidelberg (2009). https://doi.org/10.1007/978-3-642-03356-8_35

19. An, J.H., Dodis, Y., Rabin, T.: On the security of joint signature and encryption. In: Knudsen, L.R. (ed.) EUROCRYPT 2002. LNCS, vol. 2332, pp. 83–107. Springer, Heidelberg (2002). https://doi.org/10.1007/3-540-46035-7_6

Author Index

Chen, Yanping   46
Cheng, Jieren   76
Cheng, Wenjuan   29

Da, Qiaobo   76

Han, Yi-Liang   168
Huang, Ruizhang   46

Jin, Jianhua   150

Li, Chunquan   150
Li, Lian   29
Li, Ping   64
Li, Qian   76
Li, Weidong   95
Liu, Zhen   168

Meng, Yaru   46
Mo, Songsong   131

Peng, Zhiyong   131

Tian, Shan   131

Wang, Haihui   64
Wang, Junda   29
Wang, Kai   109
Wang, Liwei   131
Wang, Yong   16
Wu, Gangxiong   95
Wu, Jiadong   3

Xiao, Man   95
Xu, Weijia   46

Yang, Jian   46
Yang, Jiaoyun   29
Yang, Xiao-Yuan   168
Yang, Yue   46

Zhao, Chengye   3
Zhao, Luyao   64
Zhao, Wentao   76
Zhu, Yingxue   46

Printed in the United States
By Bookmasters

Printed in the United States
By Bookmasters